T0329360

Easy Statistics for Food Science with R

Easy Statistics for Food Science with R

Abbas F.M. Alkarkhi
Malaysian Institute of Chemical & Bioengineering Technology Universiti Kuala Lumpur, (UniKL, MICET), 78000, Melaka, Malaysia

Wasin A.A. Alqaraghuli
Skill Education Center, PA, A-07-03 Pearl Avenue, Sungai Chua, 43000 Kajang, Selangor, Malaysia

ACADEMIC PRESS
An imprint of Elsevier

Academic Press is an imprint of Elsevier
125 London Wall, London EC2Y 5AS, United Kingdom
525 B Street, Suite 1650, San Diego, CA 92101, United States
50 Hampshire Street, 5th Floor, Cambridge, MA 02139, United States
The Boulevard, Langford Lane, Kidlington, Oxford OX5 1GB, United Kingdom

Notices
Knowledge and best practice in this field are constantly changing. As new research and experience broaden our understanding, changes in research methods, professional practices, or medical treatment may become necessary.

Practitioners and researchers must always rely on their own experience and knowledge in evaluating and using any information, methods, compounds, or experiments described herein. In using such information or methods they should be mindful of their own safety and the safety of others, including parties for whom they have a professional responsibility.

To the fullest extent of the law, neither the Publisher nor the authors, contributors, or editors, assume any liability for any injury and/or damage to persons or property as a matter of products liability, negligence or otherwise, or from any use or operation of any methods, products, instructions, or ideas contained in the material herein.

British Library Cataloguing-in-Publication Data
A catalogue record for this book is available from the British Library

Library of Congress Cataloging-in-Publication Data
A catalog record for this book is available from the Library of Congress

ISBN: 978-0-12-814262-2

For Information on all Academic Press publications
visit our website at https://www.elsevier.com/books-and-journals

  Working together
to grow libraries in
developing countries

www.elsevier.com • www.bookaid.org

Publisher: Charlotte Cockle
Acquisition Editor: Megan R. Ball
Editorial Project Manager: Hilary Carr
Production Project Manager: Bharatwaj Varatharajan
Cover Designer: Victoria Pearson

Typeset by MPS Limited, Chennai, India

Dedication

Abbas

To the memory of my parents (deceased)

To my children Atheer, Hibah, and Farah

Wasin

To the memory of my father (deceased)

To my mother

Contents

Preface

Easy statistics for food science with R software was written in a simple way to introduce some statistical methods that are important to graduate students, postgraduate students, and researchers who work in the food science or food engineering fields with applications in food science. Specifically, this book was written to help researchers from different fields to analyze their data and make valid decisions. The interpretation of the results is carried out in a step-by-step manner and in an easy and clear style to enable nonstatisticians to understand and use it in their research. This book is not comprehensive because the topics were selected based on what is commonly used in the field.

The analysis is carried out step by step, and the interpretation of the results is given for each example by matching the results to the area of study where the data were obtained. The book focuses on the applications of univariate and multivariate statistical methods in the field of food science. We use real data obtained from research in the School of Industrial Technology, USM over more than 10 years of work in food science.

The development of modern statistical packages makes the analysis of data easier than before. Thus, no mathematical proofs are given in this book. Instead, the focus is on the application of statistics and correct methods for the analysis and interpretation of data.

R statistical software is used throughout the book to analyze the data. The beauty of R is that it is open source, and you can label your work based on your ideas and arrangement to produce beautiful graphs with high resolution. R is available over the internet under the General Public License (GPL online) for the Windows, Macintosh, and Linux operating systems.

Finally, we wish to thank our colleagues and friends for their continuous support, especially Dr. Yusri Yusup for his valuable comments. We would like to extend our thanks to the R software community and R family (R users and contributors to R). This book would not have been possible without the information provided online, which is easy to obtain. We thank the University of Kuala Lumpur (UnikL-MICET) for its support.

Chapter 1

Introduction

Learning outcomes

At the end of this chapter, you should be able

- *To describe multivariate analysis.*
- *To understand the benefits of multivariate methods.*
- *To arrange the results of multivariate data to be ready for analysis.*
- *To understand the difference between univariate and multivariate concepts.*
- *To know how and where to use multivariate data.*
- *To comprehend the univariate normal distribution.*
- *To comprehend the multivariate normal distribution.*

1.1 WHY SHOULD MULTIVARIATE ANALYSIS BE STUDIED?

Univariate statistical tests describe statistical methods for analyzing data related to a single variable. However, most research studies require measuring several variables for each individual or object (unit, sampling units, research units, or experimental units) in one or more samples. For example, consider assessing the flour prepared from green and ripe banana fruits based on starch, digestible starch, resistant starch, total dietary fiber, soluble dietary fiber, and insoluble dietary fiber. In this case, there are six variables measured for each sample, and these variables are usually correlated. Analyzing the results by considering one variable at a time would ignore the relationships between the variables and mislead the researcher about the real behavior of the variables in the presence of other variables. Thus, a technique that considers the relationship between different variables is required to untangle the overlapping information indicated by the correlated variables to understand the real structure of the phenomenon and the behavior of the different variables. Multivariate methods can be used to analyze data with several variables without ignoring the correlation between the variables. Multivariate analysis consists of a collection of methods that can be used for several purposes, which include identifying groups of similar individuals (objects) using cluster analysis; reducing the number of variables to a small number of indices without losing valuable information and enabling easier interpretation using principal or factor analysis; separating the data into different groups based on the measured variables using discriminant analysis; performing hypothesis testing based on several variables, such as MANOVA; and making predictions based on the relationships among the variables using multivariate multiple regression analysis.

1.2 ORGANIZATION OF MULTIVARIATE DATA

Multivariate data for several variables measured from a number of samples (items) can be presented in a table. The numbers of rows and columns of the table are specified based on the number of samples (items) and number of variables, respectively.

In general, the data for n samples (items) and k variables measured for each sample are presented in Table 1.1. In this case, the table consists of n rows to represent the samples and k columns to represent the variables measured from each sample.

Note
- The total number of samples (n) is usually greater than the number of variables measured, k.

Easy Statistics for Food Science with R. DOI: https://doi.org/10.1016/B978-0-12-814262-2.00001-7

TABLE 1.1 The Arrangement of *n* Samples and *k* Variables in Multivariate Form

Variable Sample	Y_1	Y_2	...	Y_j	...	Y_k
1	Y_{11}	Y_{12}	...	Y_{1j}	...	Y_{1k}
2	Y_{21}	Y_{22}	...	Y_{2jj}	...	Y_{2k}
⋮	⋮	⋮	...	⋮	...	⋮
i	Y_{i1}	Y_{i2}	...	Y_{ij}	...	Y_{ik}
⋮	⋮	⋮	...	⋮	...	⋮
n	Y_{n1}	Y_{n2}	...	Y_{nj}	...	Y_{nk}

1.3 EXAMPLES OF MULTIVARIATE DATA

A few examples of multivariate data may be helpful in easier demonstrating the associated calculations or illustrating concepts related to multivariate methods. Some real case studies of multivariate data related to food science will be provided. Different multivariate methods will be applied to these examples in subsequent chapters.

Example 1.1: Physicochemical properties of a banana—The physicochemical properties of banana pulp and peel flour prepared from green and ripe Cavendish bananas were investigated. The physicochemical properties, such as the pH; total soluble solids (TSS, ° Brix); water-holding capacity (WHC, g water/g dry sample); oil-holding capacity (OHC, g oil/g dry sample) at 40, 60, and 80 °C; color values (L* (lightness), a* (redness),and b* (yellowness)); back extrusion force (BEF) in N (Newtons); and viscosity were measured. The four types of flours are green peel (Gpe), ripe peel (Rpe), green pulp (Gpu), and ripe pulp (Rpu). The data are given in Table 1.2.

The objective of this study was to compare green and ripe Cavendish banana flour based on the selected physicochemical properties of the pulp and peels. Furthermore, the data were used to discriminate between the four types of flour, Gpe, Rpe, Gpu, and Rpu. The last objective was to identify the most appropriate physicochemical methods to differentiate the flours. The data will be used to identify the source of variation in the data themselves using principal components analysis (PCA).

Example 1.2: Heavy metal in cockles—A researcher wants to assess the concentration of arsenic (As) and six heavy metals (chromium (Cr), cadmium (CD), zinc (Zn), copper (Cu), lead (Pb), and mercury (Hg)) in cockles obtained from two locations. The two locations, with 20 sampling points at each location, were Kuala Juru (the Juru River) and Bukit Tambun (the Jejawi River) in the Penang State of Malaysia. The data were analyzed for the concentration of arsenic (As) and heavy metals using a graphite flame atomic absorption spectrometer (GF-AAS) for Cr, Cu, Pb, Cd, Zn, As, and a cold vapor atomic absorption spectrometer (CV-AAS) for Hg. Arsenic and heavy metals concentration were measured (mg/L) at 20 different sites. The data are given in Table 1.3.

To assess whether the cockles obtained from these rivers are polluted or not, the relationship between different parameters should be considered for more information on the behavior of each variable in the presence of other variables, which helps the researcher identify the source of the variation and use the data to discriminate between different locations. Furthermore, the contribution of each parameter in explaining the total variation in the data can be observed using multivariate methods. This study may assist in the evaluation of the impact of industry and agricultural discharge on aquaculture products in the selected area.

Example 1.3: Quality characteristics for tapioca deep-fried food—A researcher wants to investigate the quality characteristics of "tapioca" or "cassava," a popular Malaysian snack prepared by deep frying. Seven different brands of tapioca were selected based on the packaging type (polyethylene zip-lock, laminated gas packaging, and typical transparent polypropylene). Furthermore, T1 (packed in laminated pouches bag with gas packaging) was obtained from a kiosk; T2 (packed in transparent polyethylene zip-lock packaging) was acquired from an established supermarket chain; and T3, T4, T5, and T7 (all packed in transparent polypropylene plastic bag packaging) were acquired from various stores. Lastly, T6 (packed in laminated gases packaging) was obtained from a supermarket. The samples were analyzed to determine their fat oxidation (Peroxide (PV), p-anisidine (p-AV), acid value (AV) [mg KOH/g] and thiobarbituric acid (TBA) [mg MAD/kg]), fat content (%), fatty acid composition (%) (saturated fatty acid (SFA), polyunsaturated

TABLE 1.2 The Physicochemical Properties of Banana Pulp and Peel Flours Prepared From Green and Ripe Fruits

Group	pH	TSS	L*	a*	b*	WHC40	WHC60	WHC80	OHC40	OHC60	OHC80	Viscosity	Texture
Gpe	4.64	1.83	45.16	5.28	21.01	5.15	5.21	5.70	0.78	0.68	1.10	60.07	32.70
Gpe	4.32	1.73	43.10	6.03	23.02	5.16	4.87	5.97	0.78	0.77	1.06	46.90	34.16
Gpe	4.60	1.77	44.72	6.42	23.38	5.03	5.14	5.65	0.79	0.75	1.17	55.70	33.88
Gpe	4.59	1.57	37.82	5.89	21.35	5.15	4.95	6.16	0.76	0.75	1.02	49.97	37.06
Gpe	4.55	1.53	38.64	5.67	24.30	5.20	5.37	6.50	0.74	0.78	0.95	58.83	37.26
Gpe	4.33	1.67	39.85	5.36	23.73	4.99	5.53	6.00	0.80	0.77	1.02	57.53	38.02
Gpe	4.30	1.60	37.04	5.52	22.88	4.58	5.36	6.21	0.69	0.76	1.05	57.70	38.89
Gpe	5.08	1.70	36.35	4.94	21.81	5.20	5.85	5.97	0.75	0.80	1.01	46.73	39.20
Gpe	5.31	1.80	48.73	3.79	26.58	4.34	5.53	5.15	0.73	0.71	1.03	55.03	38.80
Gpe	5.33	1.80	38.42	4.38	22.66	4.97	4.81	5.71	0.85	0.75	1.02	54.53	36.90
Gpe	5.24	1.83	45.93	4.26	27.07	4.14	5.27	5.88	0.76	0.78	0.96	52.97	39.68
Gpe	5.26	1.90	34.83	4.89	21.45	5.03	4.82	5.64	0.72	0.78	0.97	54.87	40.91
Rpe	4.86	3.30	32.43	5.26	11.16	5.39	6.43	6.65	0.94	0.94	1.30	66.90	54.19
Rpe	5.31	3.20	35.61	4.77	13.10	6.10	6.56	8.10	1.06	0.98	1.33	73.47	43.99
Rpe	5.32	3.60	35.25	5.29	11.55	5.55	6.34	8.19	0.97	1.04	1.28	77.23	35.94
Rpe	5.31	3.53	32.50	5.52	11.02	6.19	6.64	8.10	0.97	0.92	1.07	79.50	55.17
Rpe	5.45	3.23	39.85	5.53	12.57	6.34	6.58	8.48	0.90	1.01	1.24	80.20	42.62
Rpe	5.69	3.47	40.31	5.55	13.07	6.21	6.72	9.26	1.01	0.96	1.39	83.90	42.62
Rpe	5.61	3.53	38.14	5.74	12.49	6.21	6.47	9.26	0.89	0.92	1.26	76.27	45.43
Rpe	5.65	3.57	39.29	5.98	12.79	6.16	6.23	8.29	1.00	0.97	1.30	79.40	47.50
Rpe	5.65	3.47	36.68	5.72	12.08	5.94	6.47	8.31	0.78	1.00	1.21	82.90	59.92
Rpe	5.68	3.50	41.08	5.32	12.97	6.17	5.90	7.92	0.96	0.98	1.37	78.30	63.24
Rpe	5.58	3.53	39.95	5.57	12.81	6.55	6.20	7.84	0.90	1.02	1.29	66.80	57.61
Rpe	5.47	3.63	40.37	6.34	14.01	6.34	5.59	7.88	0.81	1.05	1.31	72.73	59.93
Gpu	4.49	1.07	70.97	3.15	15.57	3.83	5.56	6.07	0.91	0.54	0.87	35.07	0.54
Gpu	4.37	1.17	71.70	3.41	15.22	3.72	5.52	6.03	0.91	0.57	0.83	37.00	0.63

(Continued)

TABLE 1.2 (Continued)

Group	pH	TSS	L*	a*	b*	WHC40	WHC60	WHC80	OHC40	OHC60	OHC80	Viscosity	Texture
Gpu	4.75	1.20	68.23	3.67	15.09	4.07	5.37	6.15	0.87	0.54	0.86	40.13	0.66
Gpu	5.64	1.03	64.37	3.53	14.69	4.01	5.52	6.24	0.81	0.59	0.84	38.60	0.72
Gpu	4.73	1.17	74.10	2.83	15.34	4.03	5.60	6.35	0.84	0.42	0.97	42.27	0.72
Gpu	4.59	1.13	79.25	2.49	17.69	4.08	5.78	6.53	0.77	0.47	0.85	39.13	0.70
Gpu	4.54	1.23	79.15	2.54	17.46	3.93	5.80	6.43	0.79	0.53	0.88	38.73	0.71
Gpu	5.37	1.23	78.70	1.57	16.77	3.89	5.61	6.21	0.89	0.42	0.87	40.20	0.81
Gpu	5.58	1.27	74.78	1.86	19.54	3.93	5.58	6.43	0.71	0.43	0.71	42.53	0.70
Gpu	5.65	1.37	76.62	1.68	19.29	3.74	5.79	6.43	0.73	0.45	0.79	44.67	0.68
Gpu	5.53	1.30	77.08	1.66	20.04	4.02	5.74	6.46	0.71	0.45	0.85	47.47	0.58
Gpu	5.47	1.43	75.19	1.99	21.69	3.97	5.99	6.44	0.64	0.64	0.81	45.53	0.63
Rpu	5.60	4.33	74.39	2.52	11.51	1.53	1.61	4.86	0.81	0.82	1.15	84.13	2.36
Rpu	5.57	4.03	72.00	2.88	11.85	1.50	1.74	4.04	0.82	0.84	1.03	85.17	2.49
Rpu	4.76	3.77	67.38	5.07	20.23	1.52	1.81	4.98	0.77	0.81	1.13	91.67	2.46
Rpu	5.56	4.10	70.00	3.13	12.07	1.69	1.76	5.03	0.79	0.83	1.15	87.40	2.72
Rpu	4.95	3.97	68.14	4.10	16.71	1.42	1.77	4.04	0.80	0.83	0.99	87.23	2.39
Rpu	4.84	4.40	69.82	3.67	16.11	1.28	1.56	5.01	0.69	0.87	0.94	87.33	2.39
Rpu	5.07	4.50	70.82	2.70	12.40	1.25	1.68	5.01	0.79	0.85	1.05	86.90	2.23
Rpu	4.94	4.43	72.20	2.67	12.96	1.08	1.68	4.84	0.87	0.78	1.00	90.00	2.29
Rpu	5.04	4.30	67.12	3.87	15.93	1.54	1.88	4.83	0.79	0.83	1.11	86.97	2.08
Rpu	4.93	4.57	72.69	2.59	13.31	1.18	1.58	3.99	0.70	0.73	0.95	89.73	2.20
Rpu	5.05	4.30	70.82	3.00	13.71	1.26	1.78	4.77	0.80	0.83	1.07	90.27	1.97
Rpu	5.21	4.40	74.86	2.42	12.97	1.24	1.63	4.59	0.83	0.80	1.04	87.70	2.30

TABLE 1.3 The Arsenic and Heavy Metal Contents in Cockles Obtained From 20 Sites at the Juru and Jejawi Rivers (mg/L)

Location	Cr	As	Cd	Zn	Cu	Pb	Hg
Juru	0.20	2.38	0.82	0.23	0.20	0.13	1.40
Juru	0.17	2.94	0.86	0.23	0.16	0.13	1.45
Juru	0.15	2.53	0.88	0.21	0.24	0.11	1.31
Juru	0.15	2.61	0.92	0.29	0.20	0.13	1.19
Juru	0.15	2.89	0.92	0.23	0.21	0.14	1.46
Juru	0.17	2.58	0.89	0.21	0.20	0.12	1.41
Juru	0.18	2.83	0.88	0.21	0.20	0.12	1.30
Juru	0.18	2.61	0.89	0.20	0.18	0.11	1.28
Juru	0.16	2.64	0.90	0.24	0.18	0.12	1.27
Juru	0.18	2.66	0.89	0.21	0.20	0.12	1.33
Juru	0.18	2.73	0.90	0.19	0.18	0.13	1.33
Juru	0.16	2.57	0.89	0.18	0.17	0.12	1.42
Juru	0.16	2.61	0.84	0.22	0.17	0.12	1.28
Juru	0.17	2.86	0.86	0.25	0.17	0.12	1.27
Juru	0.16	2.60	0.88	0.20	0.18	0.11	1.24
Juru	0.17	2.83	0.91	0.19	0.19	0.12	1.42
Juru	0.16	2.83	0.91	0.23	0.17	0.12	1.31
Juru	0.17	2.62	0.88	0.23	0.16	0.13	1.36
Juru	0.14	2.42	0.90	0.20	0.17	0.11	1.36
Juru	0.18	2.69	0.92	0.19	0.17	0.12	1.24
Jejawi	0.17	2.61	0.85	0.20	0.17	0.12	1.36
Jejawi	0.16	2.90	0.88	0.18	0.21	0.13	1.44
Jejawi	0.15	2.57	0.89	0.22	0.22	0.12	1.41
Jejawi	0.16	2.70	0.84	0.19	0.19	0.14	1.28
Jejawi	0.15	2.77	0.88	0.17	0.20	0.12	1.35
Jejawi	0.19	2.61	0.84	0.19	0.15	0.12	1.41
Jejawi	0.18	2.69	0.88	0.20	0.18	0.11	1.44
Jejawi	0.17	2.75	0.88	0.18	0.23	0.12	1.41
Jejawi	0.17	2.65	0.88	0.22	0.19	0.12	1.36
Jejawi	0.17	2.60	0.82	0.22	0.19	0.12	1.44
Jejawi	0.16	2.70	0.93	0.20	0.20	0.12	1.25
Jejawi	0.16	2.43	0.88	0.20	0.19	0.12	1.44
Jejawi	0.16	2.69	0.93	0.18	0.20	0.12	1.40
Jejawi	0.18	2.82	0.89	0.19	0.18	0.12	1.23
Jejawi	0.15	2.60	0.84	0.20	0.20	0.14	1.29

(Continued)

TABLE 1.3 (Continued)

Location	Cr	As	Cd	Zn	Cu	Pb	Hg
Jejawi	0.18	2.88	0.86	0.19	0.17	0.12	1.21
Jejawi	0.15	2.90	0.89	0.18	0.17	0.12	1.36
Jejawi	0.17	2.89	0.87	0.20	0.17	0.13	1.32
Jejawi	0.18	2.45	0.79	0.24	0.15	0.14	1.50
Jejawi	0.15	2.61	0.85	0.18	0.19	0.11	1.32

fatty acid (PUFA), monounsaturated fatty acid (MUFA), oil color values (lightness (L*) value, redness (a*), and yellow-ness (b*)), and moisture content (%). The data are given in Table 1.4.

Observe that the data do not provide useful information for making a comparison between different brands based on the selected parameters. The objective of the study was to assess the quality characteristics of tapioca chips regarding their fat oxidation, fat content, fatty acid composition, and moisture and oil color values. Furthermore, the objectives include finding the similarities between different brands based on the selected parameters and finding the source of the differences between these brands. The last goal was to identify the parameters responsible for the distinctions between the brands.

Example 1.4: Antioxidant in dates—Edible parts of date palm (Phoenixdactylifera) fruits (DPF) were sampled and analyzed for their antioxidant activities (AA) using the Trolox equivalent antioxidant capacity (TEAC) method, 2,2'-azinobis (3-ethylbenzothiazoline-6-sulfonic acid) radical cation ($ABTS^+$) assays and the ferric reducing/antioxidant power method (FRAP assay). The total flavonoid content (TFC) and total phenolic content (TPC) of the DPF were measured using the aluminum chloride colorimetric and Folin−Ciocalteaumethods, respectively. Four types of soft dates (SD), namely, Jiroft dates, Honey dates, Kabkab dates, and Bam dates, were used; three types of semidry dates (SDD), namely, Piarom dates, Zahedi dates, and Sahroon dates, and one type of dry date (DD) (Kharak dates) were also used. The data are given in Table 1.5.

The main goal of this study was to evaluate the AA of methanolic extracts from eight different types of DPF using the ABTS and FRAP methods. We investigated the similarities between different types of dates to study the relationships between various parameters. The goal was to understand the behavior of each variable in the presence of the other variables. Furthermore, we tested the differences among the selected varieties of dates regarding the antioxidant activities (ABTS and FRAP assay) and antioxidative compounds (flavonoids content and total phenolic). The final objective was to identify the source of the variation in the dates and to detect the effects of each parameter on the degree of differentiation among the same varieties of dates.

1.4   MULTIVARIATE NORMAL DISTRIBUTION

Most of the methods used in statistics for analyzing data are based on the normality assumption; the data are obtained from a normally distributed population. The multivariate normal distribution is an extension of the univariate normal distribution and shares many of its properties.

A brief discussion of the univariate and multivariate normal distributions is provided below.

a. Univariate normal distribution

Suppose Y is a random variable that follows the normal distribution. Then, the univariate normal distribution with mean μ and variance σ^2 is given in (1.1).

$$f(Y) = \frac{1}{\sqrt{2\pi\sigma^2}} e^{-(Y-\mu)^2/2\sigma^2} - \infty < Y < \infty \qquad (1.1)$$

It can be said that Y is distributed as $N(\mu, \sigma^2)$.

TABLE 1.4 Quality Characteristics for Tapioca Deep-fried Chips

Brand	Fat Content	Moisture	L*	a*	b*	Peroxide	p-Anisidine	Acid Value	TBA	SAF	MUFA	PUFA
T1	20.30	2.29	73.01	14.65	83.59	4.63	13.86	4.38	26.60	50.03	40.07	9.90
T1	19.96	2.24	73.00	14.63	83.55	3.45	13.91	4.48	28.10	50.38	39.82	9.80
T1	20.29	2.27	73.00	14.65	83.56	4.22	13.86	3.96	26.40	50.34	39.84	9.83
T2	18.57	3.29	23.77	44.29	40.76	17.05	10.09	24.41	79.90	49.33	27.78	8.53
T2	18.73	3.41	23.76	44.28	40.73	17.12	10.09	22.32	79.70	49.31	27.57	8.54
T2	18.44	3.33	23.75	44.28	40.70	16.97	10.01	22.20	81.10	49.81	28.10	8.47
T3	17.74	3.74	79.40	12.26	48.82	2.70	6.74	4.05	10.00	51.25	38.76	9.99
T3	17.22	3.75	79.42	12.26	48.84	2.66	6.73	3.96	9.97	51.63	38.46	9.92
T3	17.46	3.24	79.48	12.29	48.82	2.59	6.74	4.21	8.90	51.78	38.33	9.89
T4	26.05	5.32	70.05	11.45	77.16	10.73	8.53	7.97	62.40	50.77	39.33	9.90
T4	26.42	5.47	70.05	11.44	77.16	10.74	8.53	8.24	62.40	51.06	39.13	9.81
T4	26.23	5.39	70.04	11.43	77.16	10.33	8.53	8.56	57.00	50.96	39.27	9.78
T5	20.62	2.79	82.34	12.65	75.25	8.67	11.97	4.47	39.70	52.19	37.69	10.12
T5	20.89	2.62	82.34	12.63	75.20	8.55	11.96	4.20	37.80	52.33	37.72	10.05
T5	20.51	2.57	82.34	12.65	75.14	8.71	11.97	4.46	43.20	52.23	37.72	10.05
T6	29.91	4.57	67.27	41.11	114.58	10.94	2.53	3.42	7.44	51.48	37.98	10.54
T6	30.21	4.60	67.25	41.08	114.55	10.76	2.45	2.99	4.99	51.63	37.89	10.48
T6	29.92	4.55	67.24	41.08	114.57	10.98	2.65	2.91	5.11	51.70	37.87	10.43
T7	25.99	4.55	72.86	18.61	99.22	3.60	14.24	9.93	4.99	48.61	30.17	9.30
T7	26.04	4.47	72.85	18.62	99.26	2.70	15.08	8.51	4.68	49.02	30.18	9.25
T7	25.60	4.47	72.83	18.61	99.23	4.15	15.10	8.31	4.52	49.48	30.28	9.18

TABLE 1.5 Antioxidants in the Edible Parts of Date Palm Fruits

Type	TFC	TPC	AA (frap)	TEAC
Honey	1.64	2.72	11.82	23.04
Honey	1.73	2.77	11.96	21.16
Honey	1.82	2.66	10.06	22.50
Sahroon	0.96	4.83	21.08	35.82
Sahroon	0.98	4.85	18.44	35.52
Sahroon	1.63	4.73	18.87	31.88
Bam	1.64	2.23	11.04	19.43
Bam	1.77	2.36	9.66	19.43
Bam	1.96	2.10	10.02	18.00
Jiroft	0.85	1.61	7.87	25.03
Jiroft	0.89	1.62	6.96	11.82
Jiroft	1.04	1.57	7.31	13.48
Piarom	3.03	4.56	22.40	43.41
Piarom	3.29	4.39	21.67	34.60
Piarom	3.90	4.27	19.56	37.64
Kabkab	1.01	2.22	8.52	34.85
Kabkab	1.01	2.34	7.51	24.01
Kabkab	1.25	2.00	7.47	22.26
Zahedi	3.58	3.20	14.63	44.23
Zahedi	4.71	3.32	13.82	36.55
Zahedi	3.36	3.18	13.96	25.80
Kharak	55.98	117.18	329.45	404.14
Kharak	72.31	123.87	328.66	397.57
Kharak	79.58	118.21	326.30	594.71

The distribution is a bell-shaped curve, as illustrated in Fig. 1.1. Fig. 1.1 is generated using R statistical software. The codes for generating normal distribution curve are given in the Appendix.

b. Multivariate normal distribution

The multivariate normal density with mean vector μ and covariance matrix $\sum$ is given in (1.2).

$$g(Y) = \frac{1}{(2\pi)^{k/2}|\sum|^{1/2}} e^{-(Y-\mu)'\sum^{-1}(Y-\mu)/2} \tag{1.2}$$

Where k is the number of variables and $(Y-\mu)'\sum^{-1}(Y-\mu)$ is the Mahalanobis distance (statistical distance). We denote this k-dimensional normal density by $N_k(\mu, \sum)$.

Note
- Use an appropriate transformation of the data if one or more variables under study violate the normality assumption, as when the data are highly skewed, with several extreme values (high or low), called outliers, or repeated values.

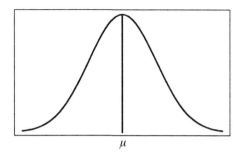

FIGURE 1.1   Normal distribution curve.

- If all the individual variables are normally distributed, then it is assumed that the joint distribution is a multivariate normal distribution.
- The real data never follow a multivariate normal distribution exactly, but the normal density is used as an approximation of the true population distribution.

FURTHER READING

Abbas, F. M. A., Foroogh, B., Liong, M. T., & Azhar, M. E. (2008). Multivariate statistical analysis of antioxidants in dates (*Phoenix dactylifera*). *International Food Research Journal, 15*, 193–200.

Alkarkhi, A. A. F. M., Ismail, N., & Easa, A. M. (2008). Assessment of arsenic and heavy metal contents in cockles (*Anadara granosa*) using multivariate statistical techniques. *Journal of Hazardous Materials, 150*, 783–789.

Alkarkhi, A. F. M., Saifullah, B. R., Yeoh, S. Y., & Azhar, M. E. (2011). Comparing physicochemical properties of banana pulp and peel flours prepared from green and ripe fruits. *Food Chemistry, 129*, 312–318.

Alqaraghuli, W. A. A., Syuzeliana, S., Yusri, Y., & Alkarkhi, A. F. M. A. An investigation of oil stability indices of fried tapioca chips in the market: Multivariate analysis.

Blogger. *R graph gallery: A collection* [Online]. Available: <http://rgraphgallery.blogspot.my/2013/04/shaded-normal-curve.html> Accessed 07.07.15.

Bryan, F. J. M. (1991). *Multivariate statistical methods : A primer*. Great Britain: Chapman & Hall.

Daniel, H. 2013. *Blog archives, high resolution figures in R* [Online]. Available: <https://www.r-bloggers.com/author/daniel-hocking/> Accessed 15.07.17.

Johnson, R. A., & Wichern, D. W. (2002). *Applied multivariate statistical analysis*. New Jersey: Prentice Hall.

Rencher, A. C. (2002). *Methods of multivariate analysis*. New York: J. Wiley.

Chapter 2

Introduction to R

Learning outcomes

At the end of this chapter, you should be able:

- *To describe R software.*
- *To describe RStudio.*
- *To know how to install R packages.*
- *To create variables, sequences, and matrices in R*
- *To apply R commands.*
- *To know how to import files into the R Environment.*
- *To write simple programs.*
- *To select appropriate R commands.*
- *To know how to plot in R.*
- *To know the working directory and set a new working directory.*
- *To understand R output and write a useful report.*
- *To import data to an R or RStudio environment.*

2.1 INTRODUCTION TO R STATISTICAL SOFTWARE

R is a software environment for statistical computing and graphical programming languages. The R language has been used by researchers at different institutions, universities, research agencies, and others. Many statistical packages are available for data analysis; however, statisticians, scientists, analysts, and others who are interested in data analysis, modeling, and producing beautiful and high-resolution graphs prefer using R software. R is made available as open-source software; under the terms of the GNU General Public License, "R is an official part of the Free Software Foundation's GNU project, and the R Foundation has similar goals to other open-source software foundations like the Apache Foundation or the GNOME Foundation." R language is analogous to the S system (language and environment) that was developed by Bell Laboratories. Millions of people use R around the world, and the number of R users increases day after day. R has become an essential, attractive, unique, and modern statistical software for the following reasons:

- R is open-source software; various sites around the world allow downloading and installing R for free, whether you are a student, academician, or affiliated with a private or public institution (agencies).
- R provides built-in functions that make analysis easy and straightforward. Data analysis in R can be accomplished by writing simple scripts to define the required parameter (variable) and calling built-in functions in the R programming language to implement the necessary operation, such as calculating the mean, standard deviation, or other statistical measures.
- R can produce beautiful and unique graphs that meet special criteria or reflect a specific idea of the work (transfer ideas to the graph).
- People without programming skills can use R easily.
- R is available for different operating systems such as Windows, Linux, MacOS, and it is easy to download and install.
- R and its library have many statistical and graphical packages for data analysis, and various calculations and graphical applications produce high-quality graphs.

Easy Statistics for Food Science with R. DOI: https://doi.org/10.1016/B978-0-12-814262-2.00002-9

- The online community is available around the world, and it is easy to interact with them, exchange ideas, and obtain help.
- R codes are available online for free; furthermore, many websites provide explanations and codes about R for free, including training, notes, and answers to questions, which other software packages do not provide.
- RStudio is a software wrapper application that provides more facilities to run R and is easier and friendlier to use than R.

This chapter covers an introduction to the R statistical software to give a starting point for people who are new (beginner) to the R language and environment. Through this chapter, readers will be able to download and install the software (R and RStudio), understand some notions and concepts used in R, and write scripts in R. We tried as much as we could to make the steps and instructions easy and understandable to everyone. Many examples are given in this chapter to guide the reader step by step and make the process enjoyable.

R statistical software and its packages can be downloaded and installed easily in a few steps, and the necessary packages associated with R software are then installed. There are many packages provided by R to implement different statistical data analysis. After installing R software, RStudio can be downloaded and installed; this is a wrapper application that can be used to run R efficiently and in a friendlier way than R.

2.2 INSTALLING R

Suppose R statistical software has not been installed yet. The reader can install and download the software for free by following the steps below.

1. The first step is to access http://ftp.heanet.ie/mirrors/cran.r-project.org/ (copy and paste the address or retype) and then click "Enter"; a new Screen will show "the Comprehensive R Archive Network" as given in Fig. 2.1.
2. It can be seen that the Screen for "The Comprehensive R Archive Network" provides three options to download R software based on the operating system of the computer, as shown below.
   1. Download R for Linux
   2. Download R for (Mac) OSX
   3. Download R for Windows

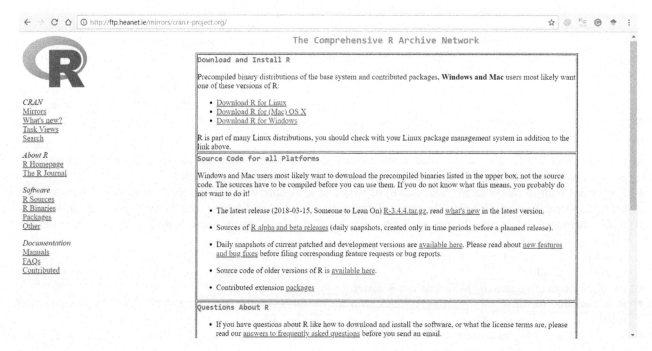

FIGURE 2.1 Comprehensive R Archive Network.

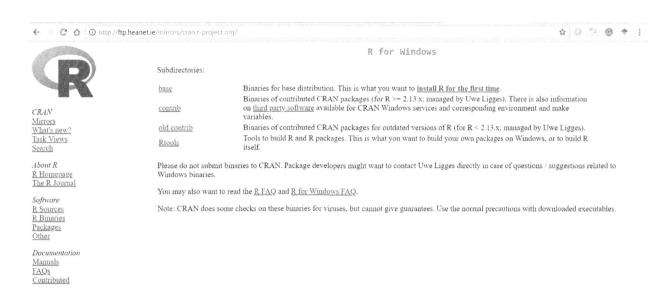

FIGURE 2.2 Showing R for Windows.

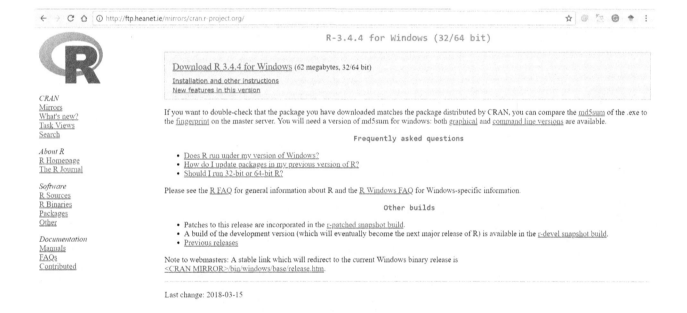

FIGURE 2.3 The screen to download R-3.4.4 for Windows (32/64 bit).

   Download R software is based on the computer operating system. If R for Windows is selected, click install R for the first time (or base) as shown in Fig. 2.2.
3. Click on the version of the software available. The latest available version (current version) is R-3.4.4 as appears on the Screen, or there may be other versions. Click on the "Download R-3.4.4 for Windows (76 megabytes, 32/64 bit)" as shown in Fig. 2.3.
4. Click on the "Download R-3.4.4 for Windows (76 megabytes, 32/64 bit)"; the file starts downloading to the computer, and in the lower bottom-left corner, there appears "R-3.4.4-win.exe" (Fig. 2.4.).
5. Once the download of R is completed, click on download file to open a new screen, and then click on "run" as shown in Fig. 2.5.The next step is to follow the instruction given to finish the installation.
6. R is already in the computer and ready to use. Double-click on the R desktop icon to start using R.

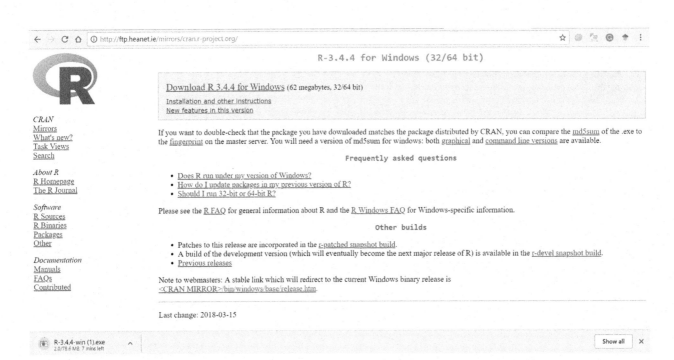

FIGURE 2.4   Showing the place of downloaded file.

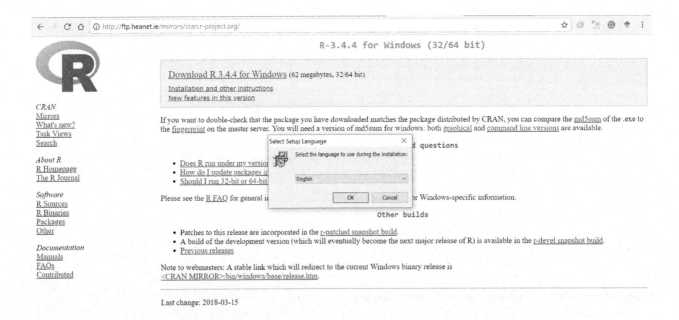

FIGURE 2.5   Showing the instructions to download the software.

2.2.1   R Documentation

It is very important to have a manual or notes to guide users using the new software, especially for beginners who are new to the R environment. R provides this service for free; it is possible to download notes and manuals using online websites such as https://www.r-project.org/other-docs.html, https://cran.r-project.org/manuals.html or other official R sites. Obtaining the documents and notes offline is possible by using the help button in the upper panel of the R software as shown in Fig. 2.6; many options are provided in the help button—one of them is Manuals (PDF).

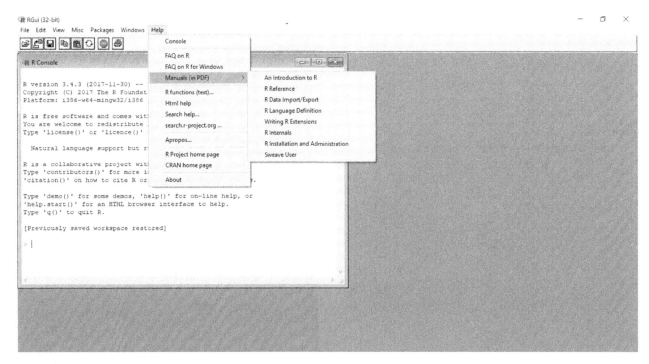

FIGURE 2.6   Showing the steps to download notes and manuals.

We found that the notes and associated documents are useful and helpful, and they provide smooth guidance, especially for beginners. Some documents have been translated into the Chinese, German, and Russian languages.

2.2.2   Installing R Packages

Many statistical packages are attached to R, and some packages are standard/base packages (built-in, loaded packages once R installation is completed). Other packages are available to users and can be downloaded from the upper panel of the R Console ("Packages"). Click "Install package(s)," and then choose the source you want to download. The last step is to choose the package you need to install from a list of Packages. The search()function can be used to show some of the loaded packages in your computer when R starts.

```
Search ()
```

The command search() will display the loaded packages.

```
> search()
  [1] ".GlobalEnv"       "package:stats" "package:graphics"
  [4] "package:grDevices" "package:utils" "package:datasets"
  [7] "package:methods"   "Autoloads"     "package:base"
```

2.3   THE R CONSOLE

Double-clicking on the R icon will open a new Screen called R Console,which is the Screen where the command should be placed; then run the command to carry out the analysis or any required calculations. At the end of the Console, there

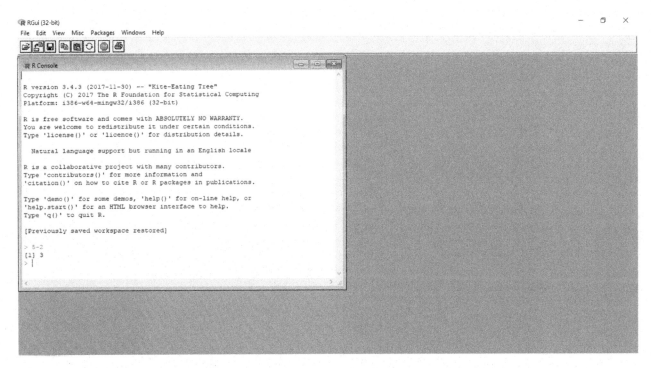

FIGURE 2.7 Showing R commands placed after the command prompt.

is a symbol >, which is called the "command prompt" by default. The R commands should be placed after the command prompt. For example, 5−2 is typed and then followed by "Enter" to get the result as shown in Fig. 2.7.

The upper panel of the Screen shows the button for File, Edit, View, MISC, Packages, Windows, and Help; each button has options to perform a certain task.

Note
- The output lines for implementing any command are preceded by [1].
- R works by calling functions to carry out the requested tasks as we have seen in the previous example for 5−2; pressing "Enter" will call the function for subtraction to run the intended job. The functions in R are called by names followed by the argument in parentheses, then press "Enter," which is a call to carry out the analysis. Usually, the functions are stored in the memory of the computer; for instance, to quit R, one should type the function q().

```
>q()
```

- Calling for help can be accomplished by typing the function help()to open a new Screen regarding the required topic. For example, the help(mean) command will provide the available information related to the mean in R library, including a description of the function, usage, arguments, references, and examples.

```
>help(mean)
```

Another command for getting help is the question mark (?), which may be used as a shortcut for calling help as in ?mean. The two commands help(mean) and ?mean are equivalent. The screen for requesting help using both commands is given in Fig. 2.8.

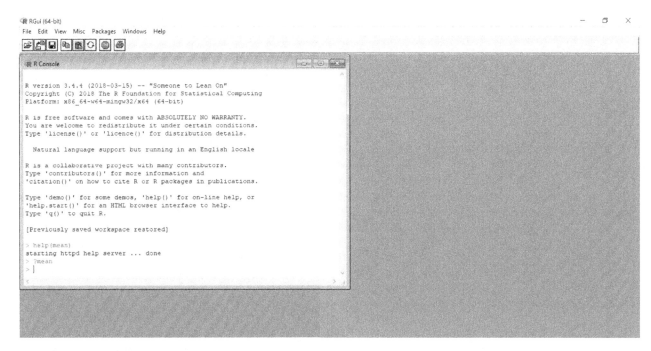

FIGURE 2.8 Showing help command.

- Typing the name of an object as a command will work similar to the function `print()`; for example, the result of A = 5−2 can be printed (shown on screen) either by typing A or typing `print(A)` as shown below.

```
>A = 5-2
>A       #Print A
[1] 3
>print(A)
[1] 3
```

- The output can be transferred to a file by using R functions such as `write.table()`, or the easiest way is to copy and paste from the R Console to a Word document.

2.4   EXPRESSION AND ASSIGNMENT IN R

Expression and assignment such as arithmetic operators, mathematical functions, and relational operators are given in this section.

1. Arithmetic operators are defined as the standard arithmetic operators, which are + , − , *, /, and ^, and each operator carries out a certain job. The function of each operator is given below.
   - + is used for addition
   - − is used for subtraction
   - * is used for multiplication
   - / is used for division
   - ^ or ** is used for exponentiation

   The precedence of arithmetic operators in R follows the standard precedence; i.e., ^ is the highest, and addition and subtraction are the lowest. Parentheses are used to control the order of the arithmetic operators.

   Example 2.1: Arithmetic operators—Calculate 3 − 1, 6÷2, and 2^3 using R.

R works as a calculator with arithmetic operators to carry out the required tasks. The result of implementing R functions for carrying out subtraction, division, and exponentiation are given below:

```
>3-1
[1] 2
>6/2
[1] 3
>2^3
[1] 8
```

2. The second concept is mathematical functions such as log (logarithm), sqrt (square root), and exp(exponential); more functions are available in R in different packages.

Example 2.2: Mathematical functions—Calculate log $(2 \div (0.5 + 0.25))$

```
>log (2/(0.5+0.25))
[1] 0.9808293
```

3. R also deals with relational operators such as $<=$, $>=$, $<$, $>$, and $!=$. The function of each operator is given below.
   - $<=$ represents less than or equal,
   - $>=$ represents greater than or equal,
   - $<$ represents less than,
   - $>$ represents greater than, and
   - $!=$ represents not equal.

2.5 VARIABLES AND VECTORS IN R

R deals with vectors as well as with a single value. The assignment operator $<-$ or equality sign $(=)$ are used to define variables (vectors or matrix) in R; for instance, $x<-3$ and $x=3$ are equivalent and give the same meaning (give the value 3 to x). The function c() is used to create or define a data set in the form of a vector.

```
>x <-c (data frame)
```

Where x represents the variable name, and numeric values (data) should be placed between the two parentheses of the function c().

Example 2.3:Create a vector—Represent the numbers 12, 3, 34, 11, 15, 9 in a vector called x.

```
>x <- c (12, 3, 34, 11, 15, 9)
```

Note
- The assignment operator $<-$ occurs side by side or acts as a "variable-defining" operator, which is the equivalent of the operator " $=$ ".
- The data in the vector must be separated by a comma (?, ?,...).
- Press "Enter" to process the command.
- R is case-sensitive; R reads Y and y as two different variables.
- If the length of two vectors X and Y is the same, then $X + Y$ in R will produce a new vector with elements representing the sum of the corresponding elements of X and Y.

Example 2.4: Two vectors of the same length—if X = 2, 7, 1 and Y = 3, 3, 5, then $X + Y$ equals to 5, 10, 6 of the same length. The R command is used to calculate $X + Y$.

```
> X = c (2, 7, 1)
> Y = c (3, 3, 5)
> X + Y
[1] 5 10 6
```

- If the length of two vectors X and Y is different, then $X + Y$ in R will repeat the shorter vector as needed to produce the new vector of $X + Y$ as shown in Example 2.5. The number of elements in the new vectors is equal to the longer vector.

Example 2.5: Two vectors of different length—if X = 2, 7 and Y = 3, 3, 5, 1, 6, then $X + Y$ equals to 5, 10, 7, 8, 8 of the same length. The R command is used to calculate $X + Y$.

```
> X = c (2, 7)
> Y = c (3, 3, 5, 1, 6)
> X + Y
[1] 5 10 7 8 8
```

It can be seen that the value 2 of the vector X is repeated three times, while the value 7 is repeated only two times to make the length of X the same as Y.

Example 2.6: Copy a vector with a zero in the middle—Create a new variable with $2n + 1$ values consisting of two copies of y with a zero in the middle. Variable y consists of three values 5, 2, and 7.

```
> y <- c (5, 2, 7)
> x <- c (y, 0, y)
> x
[1] 5 2 7 0 5 2 7
```

Note
- The function `vector name[]` is used to call a single value by locating the position.

```
> vector name [ position of the element]
```

Example 2.7: Call an element—Call the first element, and then call the third element of a vector $X = 8, 3, 11, 7$.

```
> X <- c(8, 3, 11, 7)
> X [1]
[1] 8
> X [3]
[1] 11
```

- R provides commands to call successive elements of a vector R. The colon (:) operator can be used to call successive elements of the vector.

```
Vector name [starting element : the last element]
```

Example 2.8: Call successive elements—Call the first three elements of a vector *X* given in Example 2.7. The call for the three values is X[1:3].

```
> X [1:3]
[1] 8 3 11
```

- A negative subscript associated with the calling function is used to exclude an element from a vector.

```
Vector name [- position of the value]
```

will exclude the element at that position from the vector. The command for excluding successive elements of the vector needs to specify the starting and the ending elements in the vector and place between two brackets [ ].

```
vector name [- (starting element : the last element)]
```

Example 2.9: Exclude an element or elements—Use the data given in Example 2.7 to exclude the second value and then to exclude the second and third values.

The function X[-2] is used to exclude the second value, and the function X[- (2:3)] is used to exclude the second and third values.

```
> X <- c(8, 3, 11, 7)
> X [- 2]
[1] 8 11 7
> X [- (2 : 3)]
[1] 8 7
```

- The function X[X < or X >] is used to call elements that are more or less a specific value.

```
X [ X < Specific value] or X[ X > Specific value]
```

- *Example 2.10: Relational operators*—List the elements of *X* that are less than 5, where *X* = 3, 4, 1, 7, 8, 9, 0, 4.

```
> x <- c(3, 4, 1, 7, 8, 9, 0, 4)
> x[x < 5]
[1] 3 4 1 0 4
```

2.5.1   Matrix in R

Matrix is usually used to present multivariate data, including several variables in rows and columns. So far, we have worked with scalars and how to form a single vector using R commands. The next step shows how to create a matrix from data sets using R commands.

The function matrix() is used to create a matrix in R. R can create a matrix by columns or by rows according to the given command.

```
> A = matrix (data, nrow = ?, ncol = ?, byrow = TRUE)
> A <- matrix (data, No.rows, No.columns, byrow = TRUE)
```

Example 2.11: Create a matrix—Consider that the values 3, 6, 4, 1, 8, 7 represent a variable. Create a 3-by-2 (three rows and two columns) matrix for this set of data.

The codes for defining the data set and creating the 3-by-2 matrix are given below:

```
z = c(3, 6, 4, 1, 8, 7)
z
A = matrix (z, 3, 2, byrow = FALSE)
A
```

Moreover, the result of using the function `matrix()` is

```
> z = c(3, 6, 4, 1, 8, 7)
> z
[1] 3 6 4 1 8 7
> A = matrix(z, 3, 2, byrow = FALSE)
> A
     [,1] [,2]
[1,]   3    1
[2,]   6    8
[3,]   4    7
```

This example creates the matrix by columns, which is similar to the command `A = matrix(z,3,2)`.

The output of R shows `[number, ]` and `[, number]` to represent the number of rows and the number of columns of the created matrix.

Note
- The function `matrix name[row, column]` in R provides a command to call an element of a matrix.

```
Matrix name [row, column]
```

Example 2.12: Call an element of a matrix—Call the element in row 2 and column 2 of the matrix A created in Example 2.11. The function `A[2, 2]` is used to call this element of the matrix A.

```
> A [2, 2]
[1] 8
```

- The function `matrix name[]` is used to call the entire row or column of a matrix:

```
matrix name [row number,]
```

and the command for calling a column of a matrix is:

```
matrix name [,column number]
```

Example 2.13:Call a row or a column—Call the first row and then the first column of the matrix created in Example 2.11.

```
A [1,]   #to call the first row
> A [1,]
[1] 3 1
A [,1]   #to call the first column
> A [,1]
[1] 3 6 4
```

- The matrix multiplication operator element by element is %*%.

```
%*%
```

- The solve() function is used to calculate matrix inversion.

```
solve (matrix name)
```

- The t() function is used to find the transpose of a matrix.

```
t (matrix name)
```

2.6   BASIC DEFINITIONS

1. The function seq() is used to create a sequence of values indicating the starting point, ending point, and step size.

```
seq (from = starting point, to = ending point, by = step)
```

Example 2.14:Create a sequence—Create a sequence starting from 0 to 1 (use the step equal to 0.1), and then use the step size equal to 0.5.

```
> seq (0, 1, 0.1)
[1] 0.0 0.1 0.2 0.3 0.4 0.5 0.6 0.7 0.8 0.9 1.0
> seq (0, 1, 0.5)
[1] 0.0 0.5 1.0
```

2. The rep() function is used to repeat a variable (vector) or a value.

```
> rep (variable name, times)
```

For instance, rep(2,3) means to repeat the number 2 three times.
Example 2.15:Repeating a vector—Create a vector from the data set (2, 7, 1) by repeating the data set three times.

```
> x <- c(2, 7, 1)
> rep (x, 3)
[1] 2 7 1 2 7 1 2 7 1
```

Example 2.16: Repeating elements—Repeat each element of 1:5 two times.

```
> rep (1 : 5, c(2, 2, 2, 2, 2))
[1] 1 1 2 2 3 3 4 4 5 5
or using rep (data, each = ?)
> rep (1 : 5, each = 2)
[1] 1 1 2 2 3 3 4 4 5 5
```

2.7 GRAPHS IN R

R provides great graphic facilities to produce beautiful, high-resolution graphs easily. Built-in functions are used to generate graphs; each function will carry out one task associated with a drawing graph. The plot() function is used to create a plot in R.

```
> plot ()
```

Example 2.17: Simple plot—Plot the values 7, 4, 3, 9, 2
First, define the variable, say pH, and then use the function plot().

```
> pH <- c (7, 4, 3, 9, 2)
> plot (pH)
```

The simple plot for pH values is presented in Fig. 2.9.

This graph is very simple and can be improved by adding color and lines to connect different points and characters.

Example 2.18: Colored plot—Connect the points and add color to the graph created in Example 2.17.

In Example 2.17, a very simple graph was created to represent pH values. The function type = " " is used to connect between different points, while the function col = " " is used to add the color to the graph.

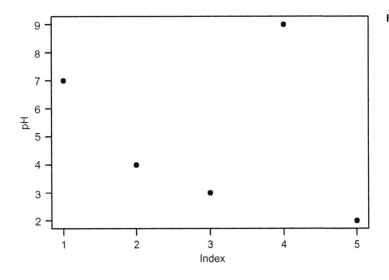

FIGURE 2.9 The plot for pH.

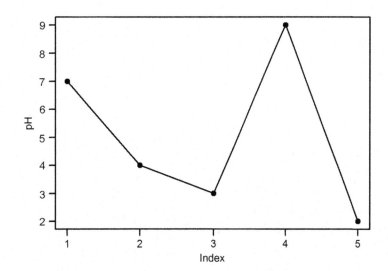

FIGURE 2.10 The plot for pH with connected points.

```
> pH <- c (7, 4, 3, 9, 2)
> plot (pH, type = "o", col = "black")
```

"o" is used to represent overplotted points and lines (there are other symbols for different descriptions). The plot with connected points and black color is presented in Fig. 2.10.

2.8   INSTALLING RSTUDIO

R statistical software version R-3.4.4 is now installed and ready for the R programming language and data analysis, as well as generating graphs and other requested calculations. RStudio is a software application that provides more facilities and is easier to work with for running R. RStudio is an IDE (Integrated Development Environment) software development application. RStudio is available for free at https://www.rstudio.com/products/RStudio/ as shown in Fig. 2.11.

Downloading RStudio starts by clicking on the "Desktop Run RStudio on your desktop" in Fig. 2.11; go to the download button as shown in Fig. 2.12.

After downloading the file, install RStudio Desktop on the computer, and then follow the subsequent instructions to complete the installation.

2.8.1   Navigate RStudio

RStudio is a user-friendly application that provides more facilities in running R commands. The RStudio screen consists of four parts. The name and the function or the purpose of each part is described below.

1. The upper-left part of the screen is called R Script, where commands are written; it is easy to correct mistakes and run the commands by clicking the run button in the upper-right corner of the top panel.
   A new R Script can be created either by clicking on the icon " + " sign on the upper-left corner of the second row or from *File > New File > R Script.*
2. The upper-right part shows the Environment (Workspace) and history, which is used to store any object, value, and function or anything that is created during the R session.
3. The R Console is the lower left of the screen where the output of running commands appears. Furthermore, commands can be written, and then "Enter" is clicked to carry out the command.

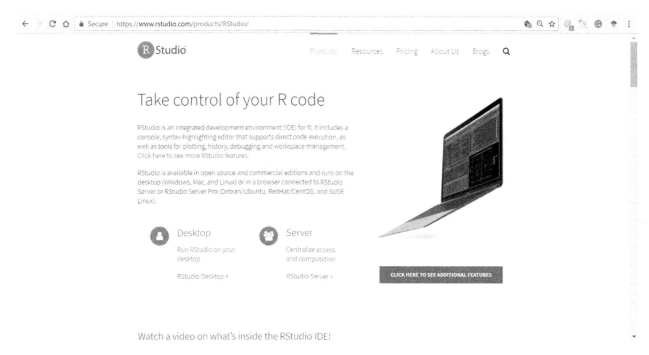

FIGURE 2.11   Installing RStudio.

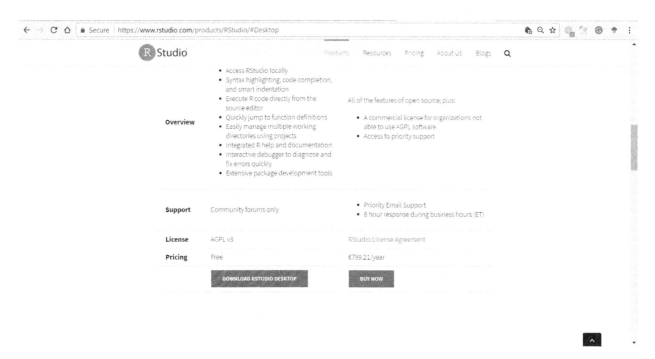

FIGURE 2.12   Showing RStudio server.

4. Files, plot, packages, Help, and Viewer. This screen is the lower-right screen. The screen shows all files and folders in the environment; the plots tab will show all created graphs, while the packages tab will allow downloading any package that is needed (Fig. 2.13).

Example 2.19: Write commands in RStudio—Use the data given in Example 2.11 to create a matrix using RStudio. The result of the created matrix is the same as was found in Example 2.11.

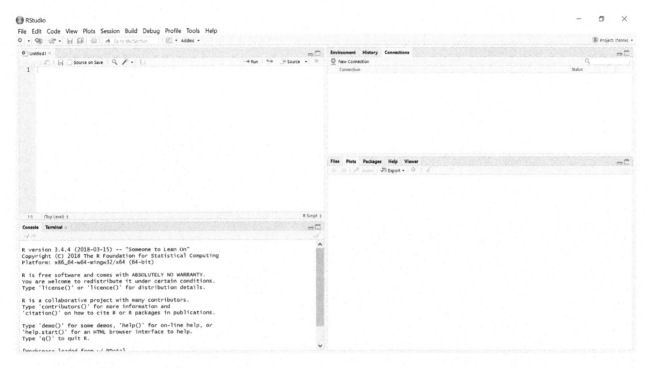

FIGURE 2.13   Showing the four parts of RStudio screen.

```
> z = c (3, 6, 4, 1, 8, 7)
> z
[1] 3 6 4 1 8 7
> A = matrix (z, 3, 2, byrow = FALSE)
> A
     [,1] [,2]
[1,]  3   1
[2,]  6   8
[3,]  4   7
```

2.9   IMPORTING DATA

R users can key in the data directly in R Console and then run the command, or the data have been stored in a file. In the case of stored data, we should import the file into the R environment. Importing files requires telling the computer the location of the stored file; this can be done if the working directory is known. The function getwd() is used to tell us the working directory of the computer. The working directory can be changed and a new working directory can be set where the file is stored by using the function setwd().

Example 2.20: Working directory—Suppose files have been stored on the Desktop; importing stored files requires setting the working directory of the computer to the location where the files have been stored.

```
> getwd ()
[1] "C:/Users/Abbas/Documents"
> setwd ("C:/Users/Abbas/Desktop")
> getwd ()
[1] "C:/Users/Abbas/Desktop"
```

R can handle several data formats, including numbers, character strings, vectors, and matrices. Researchers can store the data in a different format such as Comma Delimited, Excel, SPSS, SAS, and Stata. There is a call for each format to import the stored data. The easiest way is called a .csv (Comma delimited); this option is in Excel. This type of file can be loaded into the R environment by using the function read.csv(). The format for read.csv() is

```
read.csv(file = "file name.csv", header = TRUE)
```

The function read.csv() consists of two parts (arguments) to read the stored data. The first argument is the stored file name in .csv, and the second is the header; the header is the first row that represents the given name for the selected variables (columns). Two options can be used for the header: either FALSE if the stored data does not have column headers or TRUE if the data file has names for the variables. The data is now loaded and appears in the R Console for the R software or in the R Script for RStudio. The next step is to start the analysis of the data.

Note
- RStudio allows importing the stored data using the built-in functions on the upper-right part (Environment and Workspace or Environment and history). Click on the Environment, and then click on the Import Dataset. Five options will appear; the first option is From CSV,and the last is From Stata. Choose the option that fits the stored data, and import it to the R environment.

FURTHER READING

Chi, Y. (2009). *R Tutorial, An R Introduction to Statistics* [Online]. Available: <http://www.r-tutor.com/r-introduction> Accessed 03.05.16.

Computing, T. R. F. F. S. (2014). R: Software Development Life Cycle A Description of R's Development, Testing, Release and Maintenance Processes.

CRAN. (2016a). *The Comprehensive R Archive Network* [Online]. The R Foundation. Available: <http://ftp.heanet.ie/mirrors/cran.r-project.org/> Accessed 01.01.17.

CRAN. (2016b). *Documentation* [Online]. The R Foundation. Available: <https://www.r-project.org/other-docs.html> Accessed 20.08.17.

CRAN. *An Introduction to R* [Online]. Available: <https://cran.r-project.org/doc/manuals/R-intro.html> Accessed 19.08.17.

CRAN. *R Language Definition* [Online]. Available: <ftp://cran.r-project.org/pub/R/doc/manuals/r-release/R-lang.html> Accessed 19.08.17.

Emmanuel, P. (2005). R for Beginners France: Institute des Sciences de Montpellier II.

ENDMEMO. (2016). *R Plot Function* [Online]. Available: <http://www.endmemo.com/program/R/plot.php> Accessed 15.08.17.

Frank, M. (2006). *Producing Simple Graphs with R* [Online]. Available: <http://www.harding.edu/fmccown/r/> Accessed 05.02.16.

R Development Core Team. (2010). *R: A language and environment for statistical computing.* Vienna: Austria R Foundation for Statistical Computing.

Robert, I. K. (2014). *Accessing the Power of R* [Online]. Available: <http://www.statmethods.net/> Accessed 18.03.16.

RStudio. (2017). *RStudio, Take control of your R code* [Online]. DMCA Trademark Support ECCN. Available: <https://www.rstudio.com/products/rstudio/>.

The R Development Core Team. *The R Manuals* [Online]. Available: <https://cran.r-project.org/manuals.html> Accessed 07.08.17.

Venables, W. N., Smith, D. M., & The R Core Team. (2017). *An introduction to R-notes on R: A programming environment for data analysis and graphics.* R Core Team.

William, R. (2014). *Using R for Psychological Research: A Simple Guide to an Elegant Language* [Online]. Available: <http://personality-project.org/r/> Accessed 15.03.15.

Chapter 3

Statistical Concepts

Learning outcomes

At the end of this chapter, you should be able

- *To describe the concept of statistics.*
- *To explain the types of variables.*
- *To distinguish between population and sample.*
- *To describe the data collection.*
- *To understand the sampling techniques.*

3.1 INTRODUCTION

Statistical methods are helpful throughout the research activities, including planning, collecting data, summarizing the results, analyzing the data, and finally helping in making an intelligent decision. Researchers usually use statistics to understand the behavior of the selected parameters (variables) under study. Thus studying statistics and understanding the general concept are necessary for professionals to save time and effort and reduce costs in addition to performing forecasts. Furthermore, researchers should grasp the basics of statistics to better understand the behavior of the parameters analyzed by various statistical techniques in their field and to gain more insight into the research.

In conclusion, using the right statistical technique throughout the research will result in intelligent decisions using less time, effort, and cost.

In this chapter, we provide definitions of statistics and the two areas of statistics (descriptive and inferential); furthermore, the basic definitions of population, sample, data, and random variable are provided as well as the methods for collecting data.

3.2 DEFINITION OF STATISTICS

It is well known that researchers cannot undertake valuable research without using statistical methods for planning, analyzing, and drawing conclusions. Thus researchers should understand the definition of statistics to be able to carry out quality research and obtain great results.

Statistics has been defined in many ways by different authors. Statistics may be defined as the science of planning (conducting) studies, collecting data, organizing, summarizing, presenting, analyzing, interpreting, and extracting conclusions based on the collected data.

There are generally speaking two main areas of statistics: descriptive and inferential statistics.

Descriptive statistics can be defined as the area that describes the techniques of collecting, organizing, summarizing, and presenting the data.

Inferential statistics can be defined as the area that describes the techniques of estimation, hypothesis testing, and the study of the relationships between variables and prediction.

3.3 BASIC DEFINITIONS

Let us define some statistical terms that are useful and necessary to understand the topics in this book.

Easy Statistics for Food Science with R. DOI: https://doi.org/10.1016/B978-0-12-814262-2.00003-0

A population is a collection of items of interest (a group of concern to the researchers); the items could be humans, measurements, or other subjects under study. It can be seen in this definition that the word population can be limited— for example, types of flour (limited number—or unlimited—for example, the number of fish in a body of water.

A sample is defined as a subset (portion) of items (member, observations, subject, and individual) selected from a population. Researchers in all fields usually study samples to understand the properties of the populations. Study samples are easier than studying the whole population. Furthermore, samples are used to save time and effort and reduce costs. For example, studying the heavy metal contents in cockles can be carried out by selecting a group of cockles from a certain population (the population represents all cockles in the study area). The cockles in this group represent a sample and have characteristics similar to the cockles in the population.

Data is defined as a set of values that have been measured and collected from an experiment—for example, measurements, survey responses, and test results. Data are usually described by a random variable and based on the type of data.

Random Variable: The random variable is defined as a characteristic of interest that can assume a set of values (different values). The value of this characteristic is likely to change or vary from one trial (experiment) to another or over time. An uppercase letter is usually used to represent a variable, such as $X, Y, \ldots$, and a lowercase letter is used to represent the value of the variable.

In general, random variables can be divided into two main types:

1. Qualitative variables   2. Quantitative variables

1. Qualitative variables

    The qualitative variable is also known as an attribute or categorical variable that cannot be measured by numerical values. This type of variable takes on values that are names or labels that are not useful for calculating measures in statistics, such as the average or standard deviation and others that use numerical values. However, numbers are given when the data are being entered into the computer for identification purposes. An example of a qualitative variable is the type of flour prepared from green and ripe Cavendish bananas; four types of flour are selected, including green peel (Gpe), ripe peel (Rpe), green pulp (Gpu), and ripe pulp (Rpu). The second example involves the results of a test labeled either positive or negative, representing a qualitative variable.

2. Quantitative variables

    A quantitative variable is a measured variable (takes numerical values) and can be ranked. Measures of physicochemical properties of bananas such as pH and total soluble solids (TSS, o Brix) are examples of quantitative variables.

Two types of quantitative variables can be recognized:

1. Discrete variable   2. Continuous variable

1. Discrete variable

    A discrete variable (numeric variable) can be defined as a variable that can assume a countable or finite number of distinct values. In other words, discrete variables can take values such as 0, 1, 2, ... (there is a gap between the values) and can be counted—for example, the number of items produced in a food company every hour (day, week).

2. Continuous variable

    A continuous variable is a numeric variable that can take any value between any two specific values (interval). The values of this factor are obtained by measuring, for example, time, weight, or other readings from an instrument.

3.4   DATA COLLECTION

Researchers need information (data) to carry out research and to answer a research problem to improve or to discover a new thing. Researchers should understand the objective of the work and what type of data are needed, how to collect the data, and when to collect the data that help researchers to achieve their target and draw the right conclusions. The data required for the work depend on the type of variables to be measured and the source.

3.4.1   Methods of Data Collection

There are several ways to collect data. We can recognize four common methods of collecting data. They include:

1. Experimental approaches
2. Historical records
3. Survey approaches
4. Automated instrument approaches

Collecting data is usually carried out by using a sample, and the type of data dictates the method of selecting the items, as we will see later.

1. **Experimental approach**

   The experimental approach is the most commonly used in food science. An experiment is performed in the laboratory or field to gather the desired data that help to describe the behavior of the variable under study. For example, edible parts of date palm (*Phoenix dactylifera*) fruits (DPF) were analyzed for their antioxidant activities; this analysis requires a sample date palm fruit to be selected and then tested in the laboratory for antioxidant activities. Or a researcher may want to assess the concentration of arsenic (As) and six heavy metals (lead (Pb), chromium (Cr), cadmium (Cd), zinc (Zn), copper (Cu), and mercury (Hg)) in cockles obtained from two locations. Samples are collected from each location and analyzed for the selected parameters.

2. **Historical records**

   Historical records refer to information (studies, statistics, index) that is kept in the archives of institutions, organizations, professional associations, or any other office. The advantages of this type of data, in general, are that it is inexpensive and covers an extended period.

3. **Survey approach**

   The survey approach is a nonexperimental method that requires an original questionnaire to cover the objective of the research and all related matters; data are gathered from a community or a group of people. The number of questions should be reasonable, sufficient, and related to the subject under study. Data can be gathered in several ways, such as personal interviews, telephone interviews, mailed questionnaires, and online interviews.

4. **Automated instrument approach**

   The automated instrument approach is another way to collect data. Data are collected in this method by preparing tools to record parameters and gather the required data. This type of approach cannot be based on experimental data, for this type of data is generated without changing the input information.

3.5 SAMPLING TECHNIQUES

Most of the research is usually carried out using a portion (sample) of the population to collect information (data) about the parameters (variables) of interest. It is well known that gathering information using samples results in reduced effort, time, and cost and enables the researcher to obtain all the information about the behavior of the selected variable as the population is studied. Thus a sampling technique is defined as the procedure applied by the researchers to determine the items (observations) from the population to be included in the sample. There are several ways to select a sample from the population of interest that matches the research objective. The samples must be representative and randomly chosen to ensure independence between different items (observations). Randomization means that each sample is given the same (equal) opportunity to be selected in the sample.

The sampling techniques are as follows:

3.5.1 Simple Random Sampling

Simple random sampling is a basic technique and is considered one of the most widely used sampling techniques that gives every item (observation) in the population an equal chance of being selected in the sample. Observations in the sample are selected by giving a number to each item in the population; then, numbered cards (each card refers to a certain item) are placed in a bowl and mixed, and the required sample size is selected. Random-number tables or computer-generated random numbers can be used to select the observations of the sample easily.

Example 3.1: Final product—The final product should be checked before shipping to the market, and it should be determined whether the product is ready for market (produced items are the same; the population is homogenous). The final product in the store of a food company represents the population of interest. Testing the suitability of the product requires selecting items randomly to represent the population of the company; these items represent the sample to be tested before shipping, and a decision regarding the whole product is made based on the information provided by the sample. The product is delivered if the sample shows it is capable. Otherwise, the product should not be delivered to customers.

3.5.2 Systematic Sampling

Systematic sampling is another type of sampling technique; the members of the sample should be selected according to a random starting point chosen from N/n, where N is the population size and n is the sample size, and then every mth member is selected based on a fixed periodic interval until we reach the desired sample size.

Example 3.2: Different batches—A food company wants to select a sample of the product and test it for quality. The product in the store consists of different batches, and the sample should cover all batches produced by the company. A systematic sample should be selected; all items should be given a number in the sequence, and then a sample is selected based on the interval between different numbers. As an example, if in the store 10,000 items are produced in different batches (the product is stored according to the batches), a sample of 200 items needs to be tested. The first item should be chosen randomly from (10000/200) = 50; then, every $50mth$ ($m = 1, 2, \ldots, 200$), one item is selected. If a random start is selected from the first 50 items (between 1 to 50)—say number 3 is chosen to be the first item in the sample—then add 50 to 3 to get the second item (number 53) and so on until the last item that constitute the sample is selected. In this case, the sample would consist of the items whose number is (3, 53, 103, ...).

3.5.3 Stratified Sampling

Stratified sampling is used when the selected population is heterogeneous. Researchers select a stratified sample by separating the members of the population into distinct groups (strata) according to some essential characteristics. The members of each group are similar and different from other groups. A simple random sample will be selected from each group (stratum) to be in the required sample.

Example 3.3: Performance of a company—A new manager wants to improve the performance of a company. He suggests new regulations that help to improve performance. He wants to know what employees think (feel) about the suggested regulations. We know that different categories of employees work in a company, such asengineers, technicians, administrators, and so on. Stratified sampling should be used to represent all categories; this requires selecting a random sample from each category (stratum) to be in the required sample. The manager can study ideas from all categories by combining the selected random samples into one sample (the required sample).

3.5.4 Cluster Sampling

Cluster sampling is used when the population is divided into sections (called clusters) for geographic reasons. Cluster sampling selects clusters randomly and then uses all members in the selected clusters. This technique is different from stratified sampling since in cluster sampling we use the entire cluster, while a random sample is selected in stratified sampling.

Example 3.4: Food composition companies—There are many food composition companies distributed around Malaysia. A researcher wants to study the attitude of the employees toward a certain issue, and he is unable to survey all companies. Thus cluster sampling should be applied by selecting companies randomly, say five companies, and interviewing all the employees in the chosen companies.

FURTHER READING

Alkarkhi, A. F. M., & Low, H. C. (2012). *Elementary statistics for technologist*. Pulau Pinang, Malaysia: Universiti Sains Malaysia press.

Allan, G. B. (2007). *Elemantary statistics: A step by step approach*. McGraw-Hill.

Donald, H. S., & Robert, K. S. (2000). *A first course*. McGraw-Hill.

Mario, F. T. (2004). *Elementary statistics*. Wesky, Person Addison.

Chapter 4

Measures of Location and Dispersion

Learning outcomes

At the end of this chapter, you should be able

- *To describe measures of location.*
- *To describe measures of dispersion.*
- *To explain the covariance and correlation between variables.*
- *To plot a scatter diagram for two variables and a scatter plot matrix.*
- *To calculate and apply distances in multivariate cases.*
- *To apply R's built-in functions to calculate the measures of location and dispersion.*
- *To apply R commands beyond the built-in functions to obtain specific details or to place a specific item on a graph.*
- *To understand and interpret R output regarding measures of location and dispersion.*
- *To report on useful conclusions regarding measures of location and dispersion.*

4.1 DESCRIPTIVE STATISTICS

Researchers usually develop a general idea about the collected data by calculating some measurements to locate the center and identify the shape of the data by studying measures of center and dispersion. A quick picture about the center of a data set is achieved by using measures of the center that represent the center or middle of the data by a single value, such as the arithmetic mean, midrange, the median, and the mode, which are used to summarize data into a single number.

However, measures of center are not enough to describe a data set since no information is provided on the homogeneity of the values compared with other data sets. The spread of the values in a data set to have an idea about the homogeneity is measured using measures of dispersion (also called measures of variation).

The average or the mean of a data set is a measure used to identify the center of the data, and the variance is a measure of dispersion that shows the spread of the data around the mean. These two measures provide a clear picture of the center and the shape of the data. The relationship between different variables is of interest to the researchers; it helps them understand the behavior of each variable in the presence of other variables. Measures of location and dispersion for univariate and multivariate distributions will be given in this chapter, along with a comparison between them. Also, a correlation analysis to study the relationship between different variables, representing the bivariate data graphically using a scatter plot and the distance between two points, will be explored as well.

4.2 DESCRIPTIVE STATISTICS IN R

R statistical software provides functions to implement descriptive statistics, including measures of location and dispersion. R provides built-in functions for obtaining descriptive statistics such as the mean, median, standard deviation, variance, and others in different packages such as the psych package, Hmisc package, and pastecs package. Some of the statistical methods are part of a base installation of R and labeled *recommended*, while the user must install others.

Descriptive statistics can be derived in R using different commands to call built-in functions to implement a specific concept.

Easy Statistics for Food Science with R. DOI: https://doi.org/10.1016/B978-0-12-814262-2.00004-2

1. The function `mean()` is used to calculate the mean for a set of data in R.

```
mean (data frame)
```

2. The function `sum()` is used to calculate the sum of all of the observations (values).

```
sum (data frame)
```

3. The function `length()` is used to find the number of observations (values) in a vector.

```
length (data frame)
```

4. The function `sapply()` is used to perform the same operation (function) on a number of rows or columns of a set of data.

```
sapply (data frame, requested measure)
```

5. The function `var()` is used to calculate the variance of a dataset.

```
var (data frame)
```

6. The function `sd()` is used to calculate the standard deviation of a dataset.

```
sd (data frame)
```

7. The function `cov()` is used to calculate the covariance between two variables.

```
cov (first variable, second variable)
```

8. The function `cor()` is used to calculate Pearson's correlation between two variables.

```
cor (first variable, second variable)
```

9. The function `corr.test()` is used to calculate the coefficient of correlation and test the significance of the correlations.

```
corr.test(data frame)
```

10. The function `dist()` is used to calculate the Euclidean distance.

```
dist (data frame)
```

11. The functions `median()`, `range()`, `min()`, `max()`, and `quantile()`are used to calculate the median, range, minimum, maximum, and quantile for a variable.

```
Median ()
Range ()
Min ()
Max ()
Quantile ()
```

12. The function `pairs()` is used to create a scatter plot matrix for k variables.

```
pairs (~ Y1 + Y2 +...+ Yk, data = data frame)
```

13. The packages psych, Hmisc, and pastecs provide a summary of descriptive statistics such as the mean, median, quartiles, and others. We need to install and load the packages from their libraries, as shown below.

```
install.packages ("psych")
library (psych)
install.packages ("pastecs")
library (pastecs)
install.packages ("Hmisc")
library (Hmisc)
```

4.3 MEASURES OF LOCATION

Measures of central tendency are used to locate the center of a data set and represent the entire data set with a single value. The concept of the mean, also called the average, including the formulas for univariate and multivariate calculations, are given in the following subsections with some examples using R commands to illustrate the step-by-step analysis and interpretation of the results.

4.3.1 The Arithmetic Mean for Univariate

The mean or the average of a data set is defined as the sum of all observations (data values) divided by the total number of values (observations, subjects). The formula for calculating the mean is given in (4.1).

$$\overline{X} = \frac{\sum_{i=1}^{n} X_i}{n} = \frac{X_i + X_i + \ldots + X_n}{n} \tag{4.1}$$

where

$\overline{X}$ represents the mean,
$\sum$ (Uppercase Greek Sigma) represents the sum of all values,
X is the variable used to represent the data values, and
n represents the total number of values (observations) in the sample.

Example 4.1: Mean of the pH of green banana peels—The value of the pH of a green banana peel (grown in Malaysia) was studied as a physicochemical parameter. The data for the pH obtained from Table 1.2 for 12 different samples are listed in Table 4.1.

TABLE 4.1 The Data for the pH Values for 12 Samples of a Green Banana Peel

4.64	4.32	4.60	4.59	4.55	4.33
4.30	5.08	5.31	5.33	5.24	5.26

The mean value of the pH can easily be calculated using R statistical software. The R command to calculate the mean value of the pH using a built-in function is given below.

```
> mean (data frame)
```

The function `mean()` calculates the average of a variable using the formula given in (4.1). Because the data had been stored as a .CSV file (Example4_1), they could be loaded into the R or RStudio frame (environment) using the command read.CSV(). Alternatively, one can import the file into RStudio from the upper-right panel (Environment and then click Import Dataset) of RStudio. The mean value for the pH can be calculated in three cases using R commands as shown below.

a. The stored data (Example4_1)should be called into an R frame (environment) either by using the function read.csv ()to read the stored data or by importing the data using a built-in function. Here, the data are already loaded into the R frame, and we are ready to perform all of the required calculations.

```
mean (Example4_1 $ pH)
```

The function `mean()` is used to calculate the mean value of the pH. The character (operator) $ is used to extract a factor from a data frame, and the character # is used to add comments to a program without implementing them. The result of using built-in functions in R to calculate the mean value of the pH is given below.

```
> mean (Example4_1 $ pH)
[1] 4.795833
```

b. An alternative way to calculate the mean is to type the data in R or RStudio by writing scripts and functions as shown below.

```
pH <- c(4.64, 4.32, 4.60, 4.59, 4.55, 4.33, 4.30, 5.08, 5.31, 5.33, 5.24, 5.26)
mean (pH)
```

The result of implementing the above R commands to calculate the mean is given below.

```
> pH <- c(4.64, 4.32, 4.60, 4.59, 4.55, 4.33, 4.30, 5.08, 5.31, 5.33, 5.24, 5.26)
> mean (pH)
[1] 4.795833
```

The first command pH <- c() is used to define a variable in a vector form, pH is the variable name, <- is called the assignment operator, and c() is a function that represents the data values.

c. If more details are needed, such as the sum and number of observations, R statistical software provides space for users to implement their ideas. The R commands to calculate the sum, the number of observations, and the average are shown below.

```
pH <- c(4.64, 4.32, 4.60, 4.59, 4.55, 4.33, 4.30, 5.08, 5.31, 5.33, 5.24, 5.26)
s = sum (pH)
s                    #print s
n = length (pH)
n                    #print n
Mean = s/n
Mean                 #print Mean
```

The first row is used to define the variable pH in a vector form. Then we make a call to compute the sum of all of the pH values. The function sum() is used to calculate this sum, and the number of values is calculated by the function length(). The last formula is used to calculate the mean value of the pH (Mean = S/n). Remember that typing the objects is similar to the command print in R. The results for the sum, the number of values, and the mean value of the pH in green banana peels are shown below.

```
> pH <- c(4.64, 4.32, 4.60, 4.59, 4.55, 4.33, 4.30, 5.08, 5.31, 5.33, 5.24, 5.26)
> s = sum (pH)
> s                  #print s
[1] 57.55
> n = length (pH)
> n                  #print n
[1] 12
> Mean = s/n
> Mean               #print Mean
[1] 4.795833
```

The average value of the pH (4.795833 = 4.796) gives a general idea about the center of the data, which indicates that the value of the pH on average is 4.796 in green banana peels.

Example 4.2: Mean of the acid value (AV) in tapioca chips—A researcher wants to determine the average acid value (mg KOH/g) in tapioca chips. The acid value (AV) was determined by the AOAC method to be 940.28 (AOAC, 2000). An extracted oil sample of 0.2 g was dissolved in 10 mL of ethanol and titrated with a 0.1 M potassium hydroxide (KOH) solution using the phenolphthalein indicator until the pink color disappeared. The acid values for 21 samples for different brands are given in Table4.2.

The function mean() is used to calculate the average acid value in various tapioca chips. The average acid value can be calculated as in Example 4.1. However, only case c will be given below, and the reader can try cases a and b.

```
> AV <- c(4.38, 4.48, 3.96, 24.41, 22.32, 22.2, 4.05, 3.96, 4.21, 7.97, 8.24, 8.56, 4.47, 4.2, 4.46, 3.42,
2.99, 2.91, 9.93, 8.51, 8.31)
> mean (AV)
[1] 7.997143

# More details
> AV <- c(4.38, 4.48, 3.96, 24.41, 22.32, 22.2, 4.05, 3.96, 4.21, 7.97, 8.24, 8.56, 4.47, 4.2, 4.46, 3.42,
2.99, 2.91, 9.93, 8.51, 8.31)
> s = sum (AV)
> s                  #print s
[1] 167.94
> n = length (AV)
> n                  #print n
  [1] 21
  > Mean = s/n
  > Mean             #print Mean
  [1] 7.997143
```

TABLE 4.2 The Acid Value for Different Brands of Tapioca Chips

4.38	4.48	3.96	24.41	22.32	22.20	4.05	3.96	4.21	7.97	
8.24	8.56	4.47	4.20	4.46	3.42	2.99	2.91	9.93	8.51	8.31

In summary, the mean of the acid value for various tapioca chip samples is 7.997143. The average value provides information on the center of the data but does not give the distance of each value from the center to derive a general idea about the consistency of the values.

4.3.2 Multivariate (Mean Vector)

The concept of the mean in the case of a multivariate distribution is the same as that of the univariate distribution regarding calculation and interpretation. In the multivariate case, the average value of each variable will be placed in a column between two brackets (called a vector). The mean vector for k variables and n observations is given in (4.2), the n observation vectors are represented by $X_1, X_2, \ldots, X_n$.

$$\overline{X} = \frac{\sum X_i}{n} = \begin{bmatrix} \overline{X}_1 \\ \overline{X}_2 \\ \vdots \\ \overline{X}_k \end{bmatrix} = \frac{1}{n} X' j \tag{4.2}$$

where j is a vector of 1's, $j = \begin{bmatrix} 1 \\ 1 \\ \vdots \\ 1 \end{bmatrix}$, and $X_i = \begin{bmatrix} X_{i1} \\ X_{i2} \\ \vdots \\ X_{ik} \end{bmatrix}$

$\overline{X}$ is a $k \times 1$ vector of means, and
n represents the total number of observations associated with each variable.

Example 4.3: Heavy metals in cockles—Three heavy metal concentrations were measured in cockles (mg/L wet weight) at 10 different locations in the Juru River in the Penang state of Malaysia (Table 4.3). This dataset is part of the original data, which include seven heavy metals' concentrations.

The concentrations of chromium (Cr), arsenic (As), and cadmium (Cd) were measured from each sample. This example is a multivariate case because there are three variables (Cr, As, and Cd). Thus the mean vector for the selected variables can be calculated by computing the mean for each variable and placing the results in a vector. The mean value for Cr using (4.1) is as follows:

$$\overline{X} = \frac{\sum_{i=1}^{n} X_i}{n} = \frac{0.19 + 0.20 + \ldots + 0.15}{10} = 0.16,$$

In addition, the mean values for As and Cd are 2.67 and 0.88, respectively. The calculated means for all of the variables can be placed in a vector similar to (4.2).

The mean vector for the heavy metals in cockles can easily be calculated using the function `sapply()` in R. The command for calculating the mean value for several variables is given below.

```
sapply (data frame, requested measure)
```

Because the data for this example had been stored as a .CSV (Example4_3) file, these data could be called back into an R frame. The file can be read by using the function `read.csv()` or by importing it into RStudio using the upper-right panel (Environment and then click Import Dataset). The function `sapply()` in R is used to calculate the mean for the selected heavy metals in cockles.

TABLE 4.3 The Concentrations of Selected Heavy Metals in Cockles

Location	Cr	As	Cd
1	0.19	2.30	0.81
2	0.20	2.46	0.84
3	0.17	3.04	0.85
4	0.16	2.85	0.87
5	0.15	2.58	0.89
6	0.15	2.49	0.87
7	0.15	2.68	0.91
8	0.14	2.53	0.93
9	0.14	2.99	0.91
10	0.15	2.80	0.92

```
> Mean <- sapply (Example4_3, mean)
> Mean
   Cr     As     Cd
0.160  2.672  0.880
```

The mean vector obtained by using `sapply()` can be interpreted as follows: the values of Cr obtained from different samples fluctuated near 0.16 (i.e., these value were more or less than 0.16 but close to it), the values of As fluctuated near 2.672, and the values of Cd fluctuated near 0.88. In summary, the average concentrations of the selected heavy metals in cockles are Cr = 0.16, As = 2.67, and Cd = 0.88.

Example 4.4: Antioxidants in dates—The edible parts of date palm (*Phoenix dactylifera*) fruits (DPFs) were analyzed for their antioxidant activities (AA) using the Trolox equivalent antioxidant capacity (TEAC) method, 2,2'-azinobis (3-ethylbenzothiazoline-6-sulfonic acid) radical cation (ABTS.$^{+}$) assays, and the ferric reducing/antioxidant power method (FRAP assay). The total phenolic content (TPC) and total flavonoid content (TFC) of the DPFs were measured using the Folin−Ciocalteau and aluminum chloride colorimetric methods, respectively. The data are given in Table 4.4.

The stored data in the .CSV file `Example(4_4)` should be called first. Then we can perform the requested calculations. Similar to Example 4.3, the function `sapply()` can be used to calculate the mean vector for the antioxidant activities of the edible parts of date palm fruits.

```
> Mean <- sapply (Example4_4, mean)
> Mean
    TFC      TPC      FRAP      TEAC
10.41333  17.61625  52.46000  82.37000
```

The output of the function `sapply()` is the mean vector for the selected parameters of antioxidants in dates. The mean vector for TFC, TPC, FRAP, and TEAC appeared in the last row of the R output as follows: TFC = 10.41333, TPC = 17.61625, FRAP = 52.46000, and TEAC = 82.37000.

In summary, the values for the selected parameters of the antioxidant activities in the edible parts of date palm fruits obtained from different samples fluctuate around the corresponding calculated average values of the parameters.

TABLE 4.4 Antioxidants in the Edible Parts of Date Palm Fruits

Type	TFC	TPC	AA (frap)	TEAC
Honey	1.64	2.72	11.82	23.04
Honey	1.73	2.77	11.96	21.16
Honey	1.82	2.66	10.06	22.50
Sahroon	0.96	4.83	21.08	35.82
Sahroon	0.98	4.85	18.44	35.52
Sahroon	1.63	4.73	18.87	31.88
Bam	1.64	2.23	11.04	19.43
Bam	1.77	2.36	9.66	19.43
Bam	1.96	2.10	10.02	18.00
Jiroft	0.85	1.61	7.87	25.03
Jiroft	0.89	1.62	6.96	11.82
Jiroft	1.04	1.57	7.31	13.48
Piarom	3.03	4.56	22.40	43.41
Piarom	3.29	4.39	21.67	34.60
Piarom	3.90	4.27	19.56	37.64
Kabkab	1.01	2.22	8.52	34.85
Kabkab	1.01	2.34	7.51	24.01
Kabkab	1.25	2.00	7.47	22.26
Zahedi	3.58	3.20	14.63	44.23
Zahedi	4.71	3.32	13.82	36.55
Zahedi	3.36	3.18	13.96	25.80
Kharak	55.98	117.18	329.45	404.14
Kharak	72.31	123.87	328.66	397.57
Kharak	79.58	118.21	326.30	594.71

4.4 MEASURE OF DISPERSION (VARIATION)

Measures of dispersion are important to any set of data because the shape of the data depends on these measures. The variance and standard deviation are the most common measures of dispersion used to describe the spread of the data. The variance and standard deviation for univariate and multivariate distributions will be presented in this section.

4.4.1 Variance and Standard Deviation for Univariate

The variance of a data set is defined as the average of the squared deviations of the values from the mean. The symbol for the sample variance is S^2, and the formula for calculating the variance is given in (4.3).

$$S^2 = \frac{\sum_{i=1}^{n} \left(X_i - \overline{X}\right)^2}{n-1}$$

(4.3)

or

$$S^2 = \frac{n\left(\sum_{i=1}^{n} X_i^2\right) - \left(\sum_{i=1}^{n} X_i\right)^2}{n(n-1)}$$

The standard deviation for a sample is

$$S = \sqrt{S^2} = \sqrt{\frac{\sum_{i=1}^{n}(X_i - \overline{X})^2}{n-1}}$$

or

$$S = \sqrt{\frac{n\left(\sum_{i=1}^{n} X_i^2\right) - \left(\sum_{i=1}^{n} X_i\right)^2}{n(n-1)}}$$

Example 4.5: Variance of the pH in green banana peels—Use the data given in Example 4.1 to calculate the variance and standard deviation of the pH values measured in green banana peels (grown in Malaysia) as a physicochemical parameter.

The mean value of the pH was calculated in Example 4.1 to be 4.769. The variance and standard deviation for the pH values can be calculated using the formula given in (4.3), and they are 0.1725 and 0.41537, respectively.

The function var() in R is used to calculate the variance for a data set, and the function sd() is used to calculate the standard deviation, as shown below.

```
var (data frame )
```

The function for the standard deviation is

```
sd (data frame )
```

The data had been stored as a .CSV (Example4_5) file. The results of using built-in functions in R to calculate the variance and standard deviation for the pH in green banana peels are given below.

```
> var (Example4_5 $ pH)
[1] 0.1725356
> sd (Example4_5 $ pH)
[1] 0.4153741
```

The first result above is var(Example4_5$pH), which is the call in R to calculate the variance for the pH values, and the character $ is used to extract the pH data from the file Example4_5. The values of the variance and standard deviation that appeared in the output of R are 0.1725356 and 0.4153741, respectively. The value of the standard deviation represents the fluctuation of the data obtained from different samples around the mean value, which means that each value of the pH is above or below the mean by 0.4154. The fluctuation is low between different samples of banana peels, which reveals that the pH values in banana peels obtained from different samples are consistent.

Example 4.6: Variance of the acid value in tapioca chips—Use the data given in Example 4.2 to calculate the variance and standard deviation for the acid values of tapioca chips as a chemical test.

The data had been stored as a .CSV (Example4_6) file. As in Example 4.5, the results of using R functions to calculate the variance and standard deviation for the acid values of tapioca chips are given below.

```
> var (Example4_6 $ AV)
[1] 43.99826
> sd (Example4_6 $ AV)
[1] 6.633119
```

The results showed that the value of the standard deviation is 6.633119, which represents the fluctuation of the data around the mean value. The fluctuation of the acid value is high, which indicates that the differences in the acid values between different brands are high. The fluctuation in the acid values could be due to the material used, the packaging technology, or another reason that cannot be properly controlled for.

4.5 COVARIANCE

It is important to know the association between different variables to have a clear picture of the behavior of the variables under study. The sample covariance measures the linear association between two variables. When two variables are measured for each research unit, we have a bivariate case. Consider two variables X and Y; a measure of the linear association between the measurements of two variables X and Y is provided by using (4.4) to calculate the sample covariance. The symbol for the sample covariance is S_{XY}. The formula for calculating the covariance between any two variables is given in (4.4).

$$S_{XY} = \frac{1}{n-1}\sum_{i=1}^{n}(X_i - \overline{X})(Y_i - \overline{Y}) \tag{4.4}$$

- S_{XY} will be positive if increasing the values of one variable would result in increasing the values of the other variable, or vice versa.
- S_{XY} will be negative if increasing the values of one variable would decrease the values of the other variable.
- S_{XY} will be approximately zero if there is no association between the values of the two variables.

4.5.1 Covariance Matrices (Multivariate)

In the case of several variables, the covariance between all the possible combinations of different variables can be placed between two brackets called a matrix.

The sample covariance matrix $S = (S_{ij})$ is the matrix of variances and covariances of the k variables, as given in (4.5).

$$S = (S_{ij}) \begin{bmatrix} S_{11} & S_{12} & \cdots & S_{1k} \\ S_{21} & S_{22} & \cdots & S_{2k} \\ \vdots & \vdots & & \vdots \\ S_{k1} & S_{k2} & \cdots & S_{kk} \end{bmatrix} \tag{4.5}$$

The covariance matrix is a symmetric matrix; the main diagonal represents the variances between different variables, and the off-diagonal represents the covariances.

- The variance S_i^2 is represented by S_{ii} in the covariance matrix; S_{11} represents the variance of the first variable, while S_{ij} represents the covariance. For instance, S_{12} represents the covariance between variable 1 and variable 2.

4.6 CORRELATION

The correlation is a statistical measure used to determine whether the relationship between two variables exists or not. The correlation is used to measure the strength and the direction of the linear relationship between two variables.

The correlation coefficient (Pearson's product-moment correlation coefficient) is used to measure the linear relationship between two quantitative variables; the value does not depend on the scale of measurement. The correlation between two variables is defined in (4.6).

$$r_{XY} = \frac{S_{XY}}{S_X S_Y} = \frac{\sum_{i=1}^{n}(X_i - \overline{X})(Y_i - \overline{Y})}{\sqrt{\sum_{i=1}^{n}(X_i - \overline{X})^2 \sum_{i=1}^{n}(Y_i - \overline{Y})^2}} \tag{4.6}$$

The numerator of (4.6) represents the covariance between the two variables; the denominator represents the product of the variance of the two variables.

Note
- The value of r must be between -1 and 1.
- r measures the strength of the linear association between two variables.
- If $r = 0$, this implies a lack of linear association between the two variables (no linear relationship).
- The sign of r indicates the direction of the association.

4.6.1   Correlation Matrices

In the case of k variables, the correlation between variables for all the possible combinations of k variables can be arranged in a matrix called a correlation matrix. The sample correlation matrix is given in (4.7).

$$R = (r_{ij}) = \begin{bmatrix} 1 & r_{12} & \cdots & r_{ik} \\ r_{21} & 1 & \cdots & r_{2k} \\ \vdots & \vdots & & \vdots \\ r_{k1} & r_{k2} & \cdots & 1 \end{bmatrix} \tag{4.7}$$

- The sample correlation matrix is analogous to the covariance matrix with correlations in place of covariances.

Example 4.7: Chemical parameters for fish crackers—Seven different brands of fish crackers were selected based on the packaging type. The samples were analyzed for the chemical test parameters; the parameters are the peroxide (PV), p-anisidine (PAV), acid value (AV [mg KOH/g]), and quantity of thiobarbituric acid (TBA [mg MAD/kg]). The data are given in Table 4.5.

The covariance matrix and the correlation matrix for the chemical parameters of fish crackers can easily be calculated by the equations defined in(4.4) and (4.6), respectively.

The commands for calculating the variance−covariance and correlation matrices between different variables are demonstrated and presented in an easy way.

The function cov() is used to calculate the covariance matrix between different chemical parameters (as defined in (4.4)) and to place the results in a matrix form.

```
cov (first variable, second variable)
```

The data for the chemical parameters of fish crackers had been stored as a .CSV (Example4_7) file.

```
cov (Example4_7)
```

The results of using R commands to produce a variance−covariance matrix for the chemical parameters are given below.

```
round (data frame, digits = Number of digits)
```

```
> round (cov(Example4_7), digits = 2)
        PV     PAV    AV    TBA
PV    0.25    1.59  0.08   0.03
PAV   1.59   43.24  1.17  14.54
AV    0.08    1.17  0.19   0.29
TBA   0.03   14.54  0.29   8.21
```

TABLE 4.5 The Results of the Selected Parameters for a Chemical Test of Fish Crackers

AV	PAV	AV	TBA
1.50	24.41	3.77	11.38
1.48	23.94	3.75	11.48
1.51	24.36	3.83	11.51
1.39	24.12	4.00	9.00
1.31	23.38	3.98	8.64
1.28	23.81	3.92	8.95
0.61	6.70	2.99	6.52
0.60	7.07	3.02	6.46
0.60	7.05	3.16	6.50
0.60	13.07	2.83	5.63
0.59	13.52	2.82	5.70
0.59	13.46	2.83	5.47
1.31	9.86	3.94	2.81
1.21	9.38	3.95	2.58
1.20	9.48	3.84	2.54
1.60	20.64	3.08	7.42
1.61	20.82	3.09	7.97
1.59	21.83	3.18	7.44
1.99	14.16	3.22	3.42
2.09	14.16	3.23	3.83
2.07	14.23	3.27	3.98

The first command is the function round(), which is used to round the numbers to the specified number of digits. The results for the chemical parameters are rounded up to two digits.

The results of applying R commands show that whereas the main diagonal represents the variance of each parameter (0.25, 43.24, 0.19, and 8.21), the off-diagonal values represent the covariances between the different parameters; for instance, the covariance between peroxide (PV) and thiobarbituric acid (TBA) is 0.03.

The correlation matrix between different variables can be calculated as defined in (4.6). In R, the built-in function cor() can be used to compute the correlation matrix between different chemical parameters for fish crackers and to place the results in a matrix form. The results between all possible combinations of the relationship (correlations) for the chemical parameters of fish crackers are given below.

```
cor(Example4_7)
```

The results of applying the function cor() are as follows:

```
> round(cor(Example4_7), digits = 2)
       PV   PAV   AV   TBA
PV   1.00  0.48  0.34  0.02
PAV  0.48  1.00  0.40  0.77
AV   0.34  0.40  1.00  0.23
TBA  0.02  0.77  0.23  1.00
```

The correlation matrix produced by R shows that the main diagonal is 1 for all of the parameters, and the off-diagonal values represent the correlations between all possible combinations of different chemical parameters. For instance, the correlation between peroxide (PV) and thiobarbituric acid (TBA) is 0.02.

Note
- R provides another command to compute the correlations between different variables, and the results will be similar to the first command.

```
cor (data frame, use = "complete.obs")
```

- The correlations between different chemical parameters have been computed without providing information on whether the correlation is significantly different from zero (i.e., whether the relationship between different parameters is true and exists) or not. Testing the significance of the correlation requires using another function in R: the function corr.test(). Thus we also use this function to test the significance of the correlations and to compute the correlations.

However, the command corr.test() for testing the significance of the correlations needs to install and load two packages: psych and GPA rotation. This command will give the correlation matrix and the significance levels (probability values) of the correlations for all of the possible combinations of two variables.

```
install.packages("psych")
library(psych)
install.packages("GPArotation")
library(GPArotation)
corr.test(Example4_7)
```

The function corr.test() was used to test the significance of the relationship for the chemical parameters of fish crackers.

```
> corr.test(Example4_7)
Call:corr.test(x = Example4_7)
Correlation matrix
        PV    PAV    AV    TBA
PV    1.00   0.48   0.34   0.02
PAV   0.48   1.00   0.40   0.77
AV    0.34   0.40   1.00   0.23
TBA   0.02   0.77   0.23   1.00
Sample Size
[1] 21
Probability values (Entries above the diagonal are adjusted for multiple tests.)
        PV    PAV    AV    TBA
PV    0.00   0.13   0.39   0.92
PAV   0.03   0.00   0.28   0.00
AV    0.13   0.07   0.00   0.65
TBA   0.92   0.00   0.32   0.00
To see confidence intervals of the correlations, print with the short = FALSE option
```

The first part of the results produced by using the function corr.test() is the correlation matrix between different chemical parameters for fish crackers, and the second part of the results is a table of probability values that represents the significance level (alpha). The significance levels of different correlations can easily be extracted from the table of probability values, which depends on the lower-triangular values (i.e., the values below the diagonal). For instance, the first column gives the significance level between PV and the other variables, and the second column gives the significance level between p-anisidine (PAV) and the other variables excluding peroxide, and similarly for the other columns.

The probability values in the first column indicated a strong positive relationship between PV and PAV (p-value < 0.03). Furthermore, there was a positive correlation between PAV and TBA (p-value < 0.00), while the correlation between the PAV and AV values was significant at p-value < 0.07. The relationship between the other parameters did not show significant relationships, as the p-values are high. The relationship between the PAV and the other parameters indicates that the PAV has a strong association with all of the parameters and shares a common origin with them. Other correlations can be explained in the same manner.

Example 4.8: Heavy metals in cockles—The concentrations of arsenic (As) and six heavy metals (chromium (Cr), cadmium (Cd), zinc (Zn), copper (Cu), lead (Pb), and mercury (Hg)) in cockles obtained from two locations were studied. These two locations, with 20 sampling points at each location, were Kuala Juru (the Juru River) and Bukit Tambun (the Jejawi River) in the Penang State of Malaysia. The data were analyzed for the concentrations of arsenic (As) and heavy metals using a graphite flame atomic absorption spectrometer (GF-AAS) for Cr, Cd, Zn, Cu, Pb, and As and a cold vapor atomic absorption spectrometer (CV-AAS) for Hg. The arsenic and heavy metals concentrations were measured (mg/L) at 20 different sites. The data for the selected heavy metals (six heavy metals) and arsenic are given in Table 1.3.

The data had been stored as a .CSV(Example4_8) file. The results for the variance—covariance matrix for the selected heavy metals in cockles are calculated as defined in (4.5) and rounded up to three digits. The calculations were carried out using the function cov() in R.

```
> round(cov(Example4_8),digits=3)
    Cr    As     Cd     Zn Cu Pb     Hg
Cr  0 0.000  0.000  0.000  0  0  0.000
As  0 0.021  0.001 -0.001  0  0 -0.002
Cd  0 0.001  0.001  0.000  0  0 -0.001
Zn  0 -0.001 0.000  0.001  0  0  0.000
Cu  0 0.000  0.000  0.000  0  0  0.000
Pb  0 0.000  0.000  0.000  0  0  0.000
Hg  0 -0.002 -0.001 0.000  0  0  0.006
```

The correlations between all possible combinations of the different parameters are calculated as defined in (4.6). The correlation matrix between the different heavy metals and arsenic parameters are produced by using the function cor() in R.

```
> round(cor(Example4_8),digits=3)
       Cr     As     Cd     Zn     Cu     Pb     Hg
Cr  1.000  0.007 -0.321  0.009 -0.317  0.056  0.071
As  0.007  1.000  0.295 -0.164 -0.049  0.162 -0.133
Cd -0.321  0.295  1.000 -0.078  0.276 -0.266 -0.243
Zn  0.009 -0.164 -0.078  1.000 -0.079  0.282 -0.110
Cu -0.317 -0.049  0.276 -0.079  1.000 -0.056  0.035
Pb  0.056  0.162 -0.266  0.282 -0.056  1.000  0.190
Hg  0.071 -0.133 -0.243 -0.110  0.035  0.190  1.000
```

The correlation coefficients for all possible combinations with p-values to test the significance of the correlations between the heavy metals and arsenic parameters are provided by the function corr.test().

```
> corr.test (Example4_8)
Call:corr.test(x = Example4_8)
Correlation matrix
      Cr    As    Cd    Zn    Cu    Pb    Hg
Cr  1.00  0.01 -0.32  0.01 -0.32  0.06  0.07
As  0.01  1.00  0.29 -0.16 -0.05  0.16 -0.13
Cd -0.32  0.29  1.00 -0.08  0.28 -0.27 -0.24
Zn  0.01 -0.16 -0.08  1.00 -0.08  0.28 -0.11
Cu -0.32 -0.05  0.28 -0.08  1.00 -0.06  0.03
Pb  0.06  0.16 -0.27  0.28 -0.06  1.00  0.19
Hg  0.07 -0.13 -0.24 -0.11  0.03  0.19  1.00
Sample Size
[1] 40
Probability values (Entries above the diagonal are adjusted for multiple tests.)
      Cr    As    Cd    Zn    Cu   Pb Hg
Cr  0.00  1.00  0.92  1.00  0.93  1.001
As  0.97  0.00  1.00  1.00  1.00  1.001
Cd  0.04  0.06  0.00  1.00  1.00  1.001
Zn  0.96  0.31  0.63  0.00  1.00  1.001
Cu  0.05  0.77  0.08  0.63  0.00  1.001
Pb  0.73  0.32  0.10  0.08  0.73  0.001
Hg  0.66  0.41  0.13  0.50  0.83  0.240
To see confidence intervals of the correlations, print with the short = FALSE option
```

The probability values in the first column indicated a strong positive relationship between chromium (Cr) and cadmium(Cd) (p-value < 0.04) and between chromium and copper (Cu) (p-value < 0.05). Moreover, there were positive relationships between arsenic (As) and Cd (p-value < 0.06), between Cd and Cu (p-value < 0.08),and between Zn and lead(Pb) (p-value < 0.08). The results of the other heavy metal parameters did not show significant relationships.

Note
- The function `cor.test()` can be used to test the correlations between two variables as an alternative R command. It works simply by nominating the two variables X and Y.

```
cor.test (x,y)
```

4.7 SCATTER PLOT

A scatter diagram is a graph of paired (X, Y) data values. It consists of a horizontal axis to represent the range of one variable and a vertical axis to represent the range of the second variable. A scatter diagram can provide a general picture of the relationship between two variables.

Example 4.9: Scatter diagram for the acid value and p-anisidine—Use the data in Table 4.6 for the PAV and acid values (AV) in banana chips, which were computed by a chemical test, to construct a scatter diagram.

R statistical software provides functions to generate different plots for various statistical purposes. Furthermore, R allows one to place a specific item on any generated graph, e.g., a color can be chosen and changed, as well as the shape of the marks or the size marks, the legend, and other things related to graphs. Moreover, text, points, and lines can be placed in the plot. The function `plot()` is used to generate different graphs.

```
plot (x,y)
```

The data for the p-anisidine and acid values had been stored as a `.CSV(Example4_9)` file. The function `plot()`is used to generate a scatter diagram for the p-anisidine and acid value.

TABLE 4.6 The Results for the p-Anisidine and Acid Values in Banana Chips

p-Anisidine	22.84	22.77	22.81	2.56	2.51	2.55	18.69	18.49	18.54	7.08	
Acid value	4.03	4.03	3.98	2.35	2.28	2.34	3.67	3.65	3.68	2.46	
p-Anisidine	6.98	7.04	10.69	10.62	10.68	10.39	10.38	10.36	3.03	2.96	2.92
Acid value	2.49	2.47	3.12	3.12	3.12	4.26	4.25	4.21	2.17	2.20	2.18

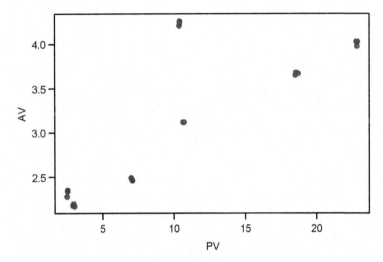

FIGURE 4.1   Scatter plot of the p-anisidine versus the acid value showing a positive relationship.

```
plot (Example4_9 $ PV, Example4_9 $ AV, xlab = "", ylab = "",
      cex = 0.5, pch = 10, col = "blue", cex.axis = 0.5, xaxt = "n", yaxt = "n")
```

Where `ylab = ""` represents the y-axis and `xlab = ""` represents the x-axis. The command `pch = ?` is used to specify the symbols to be used for plotting points. The function `col = ""` is used to specify the color.

The scatter diagram for the acid value and p-anisidine is given in Fig. 4.1. This figure reveals that there is a clear trend for the relationship between the p-anisidine and the acid value. Moreover, a positive relationship is observed between them because increasing the value of the p-anisidine would increase the acid value. This observation comprises a general conclusion about the behavior of p-anisidine and the acid value. However, the strength of the relationship cannot be revealed by a scatter diagram and should be studied through the correlation coefficient.

Note
- Scatter plots can have different patterns than the one shown in Fig. 4.1. Other types of scatter plots are shown in Fig. 4.2. Using the same code for a scatter plot with different data sets can produce the scatter plots illustrated in Fig. 4.2.
a. There is no relationship between X and Y.
b. There is a nonlinear relationship between X and Y.

4.7.1   The Scatter-plot Matrix

The scatter-plot matrix is used when there are several variables under study. It is a nice way to present and organize scatter plots for all possible pairs of variables to be within one view. R statistical software provides functions for creating a scatter plot matrix in different packages. The function `pairs()` is usually used to create a scatter plot matrix.

```
Pairs ( ~ Y1 + Y2 + ...+ Yk, data = data frame)
```

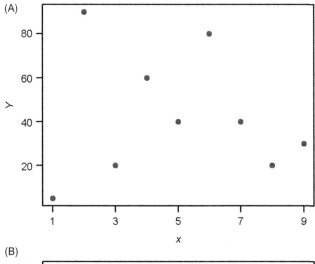

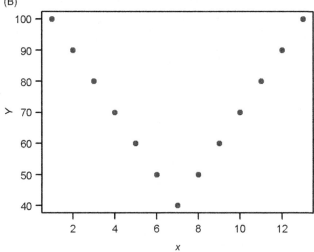

FIGURE 4.2 Scatter plots for different types of relationships: (A) no relationship between X and Y, (B) a nonlinear relationship between X and Y.

Example 4.10: Scatter plot matrix for the chemical parameters of deep fried snacks—Build a scatter plot matrix for the chemical test of four parameters obtained from two types of deep-fried snacks, namely, muruku and tapioca. The data are given in Table 4.7.

The function `pairs()` can be used to create a scatter plot for all of the pairs.

```
pairs(~ PV + PAV + AV + TBA, data = Example4_10, col = "blue", cex = 0.5, pch = 10,cex.axis = 0.7,
tck = -0.04, mgp = c(3, .3, 0))
```

The results of applying the function `pairs()` is the scatter plot matrix shown in Fig. 4.3. Observe that the scatter plot matrix includes scatter plots for all possible combinations of the different chemical parameters obtained from two deep-fried snacks.

4.8 DISTANCE

The straight line between two points A and B with coordinates $A = (X_1, X_2, \ldots, X_k)$ and $B = (Y_1, Y_2, \ldots, Y_k)$ is called the Euclidean distance, as defined in (4.8).

$$d = \sqrt{(X_1 - Y_1)^2 + (X_2 - Y_2)^2 + \ldots + (X_k - Y_k)^2}$$

(4.8)

TABLE 4.7 The Results of a Chemical Test for Selected Deep-fried Snacks

Product	PV	PAV	AV	TBA
Muruku	3.73	14.03	6.63	0.04
Muruku	3.24	14.04	8.38	0.04
Muruku	3.59	14.05	9.85	0.04
Muruku	10.31	3.69	17.15	0.02
Muruku	10.26	3.66	15.68	0.02
Muruku	10.23	3.66	14.37	0.02
Muruku	6.03	8.37	4.29	0.03
Muruku	6.03	8.37	5.02	0.03
Muruku	5.98	8.39	4.44	0.03
Tapioca	4.63	13.86	4.38	26.60
Tapioca	3.45	13.91	4.48	28.10
Tapioca	4.22	13.86	3.96	26.40
Tapioca	17.05	10.09	24.41	79.90
Tapioca	17.12	10.09	22.32	79.70
Tapioca	16.97	10.01	22.20	81.10
Tapioca	2.70	6.74	4.05	10.00
Tapioca	2.66	6.73	3.96	9.97
Tapioca	2.59	6.74	4.21	8.90

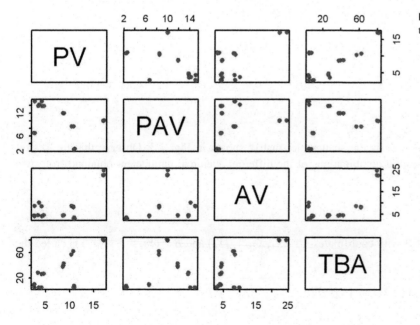

FIGURE 4.3 A scatter plot matrix for a chemical test of fish crackers.

Each coordinate contributes equally to the calculation of (4.8). This property is undesirable for most statistical purposes and is also uninformative. Thus developing a formula that takes into account the differences in the variations of the variables, as well as the correlation or covariances, is desirable. In this case, the contribution depends on the variation and the relationship between different variables in the presence of other variables.

A statistical distance is defined as the distance that accounts for the variances of the variables and their covariances or correlations, i.e., the statistical distance takes into account the variation and the correlation (or covariances) between different variables. The formula for the statistical distance is given in (4.9).

Let the points A and B have coordinates such that $A = (X_1, X_2, \ldots, X_k)$ and $B = (Y_1, Y_2, \ldots, Y_k)$; then the statistical distance is given by (4.9).

$$\sqrt{\frac{(X_1 - Y_1)^2}{S_{11}} + \frac{(X_2 - Y_2)^2}{S_{22}} + \ldots + \frac{(X_k - Y_k)^2}{S_{kk}}} \tag{4.9}$$

Where S_{ii} is the sample's variance.

Note
- If a random variable has a much larger variance than the others, it will contribute less to the squared distance (it receives less weight).
- Two highly correlated random variables will contribute less than two variables that are less correlated.
- The difference between the Euclidean distance and statistical difference is divided by the standard deviation in the statistical formula.

Example 4.11: Distances for heavy metals in cockles—Calculate the distances for the data of heavy metals and arsenic in cockles used in Example 4.8. The data consist of the first 10 samples obtained from Table 1.3 and 7 parameters (6 heavy metals and arsenic) measured from each sample.

The built-in function dist() in R can be used to calculate the distances between all sampling points and produce a distance matrix for all heavy metals and arsenic parameters. The data had been stored as a .CSV (Example4_11) file. The distance matrix for arsenic and heavy metals in cockles is generated by the function dist() in R.

```
round(dist(Example4_11), digits = 2)
```

The results of applying the function dist() for the heavy metals and arsenic parameters are rounded up to two digits.

```
> round(dist(Example4_11),digits=2)
        1     2     3     4     5     6     7     8     9
2    0.57
3    0.20  0.44
4    0.34  0.43  0.18
5    0.53  0.10  0.39  0.39
6    0.22  0.37  0.12  0.24  0.32
7    0.47  0.19  0.30  0.26  0.18  0.27
8    0.27  0.37  0.11  0.14  0.34  0.14  0.22
9    0.31  0.35  0.14  0.10  0.32  0.16  0.20  0.06
10   0.30  0.31  0.14  0.17  0.27  0.11  0.17  0.07  0.08
```

Example 4.12: Distances for the chemical parameters for deep-fried snacks—Calculate the distance for the data used in Example 4.10 (Table4.7). The data consist of samples for the two types of deep-fried snacks (muruku and tapioca) and four parameters measured for each sample.

The built-in function dist() in R can be used to calculate the distances between all of the selected chemical tests and produce a matrix of these distances. The data had been stored as a .CSV (Example4_12) file. The entries of the distance matrix for the chemical test for selected deep-fried snacks are rounded up to two digits.

```
> round (dist (Example4_12), digits = 1)
        1     2     3     4     5     6     7     8     9    10    11    12    13    14    15    16    17
2     1.8
3     3.2   1.5
4    16.2  15.3  14.3
5    15.2  14.5  13.7   1.5
6    14.5  13.9  13.1   2.8   1.3
7     6.5   7.5   8.3  14.3  13.0  11.9
8     6.3   7.2   7.8  13.7  12.4  11.3   0.7
9     6.5   7.4   8.2  14.2  12.9  11.8   0.2   0.6
10   26.7  26.9  27.1  31.7  31.1  30.7  27.2  27.2  27.2
11   28.1  28.3  28.6  33.2  32.6  32.2  28.7  28.7  28.7   1.9
12   26.5  26.7  27.0  31.8  31.2  30.7  27.0  27.0  27.0   0.6   1.9
13   83.0  82.7  82.4  80.7  80.9  81.1  83.1  82.9  83.1  58.4  57.3  58.8
14   82.4  82.1  81.9  80.4  80.5  80.6  82.5  82.3  82.4  57.5  56.4  58.0   2.1
15   83.7  83.5  83.2  81.8  81.9  82.0  83.8  83.6  83.7  58.8  57.6  59.2   2.5   1.4
16   12.7  13.1  13.7  18.4  17.4  16.5  10.6  10.7  10.6  18.2  19.5  17.9  74.3  73.6  74.8
17   12.7  13.1  13.7  18.5  17.4  16.6  10.6  10.7  10.6  18.2  19.5  18.0  74.3  73.6  74.9   0.1
18   11.8  12.2  12.8  17.8  16.7  15.8   9.7   9.7   9.6  19.2  20.5  19.0  75.3  74.6  75.9   1.1   1.1
```

REFERENCE

AOAC. (2000). *Official methods of analysis of AOAC international* (17th ed.). United States of America: AOAC International.

FURTHER READING

Abbas, F. M. A., Foroogh, B., Liong, M. T., & Azhar, M. E. (2008). Multivariate statistical analysis of antioxidants in dates (*Phoenix dactylifera*). *International Food Research Journal, 15*, 193–200.

Alkarkhi, A. F. M., Ismail, N., & Easa, A. M. (2008). Assessment of arsenic and heavy metal contents in cockles (*Anadara granosa*) using multivariate statistical techniques. *Journal of Hazardous Materials, 150*, 783–789.

Alkarkhi, A. F. M., & Low, H. C. (2012). *Elementary statistics for technologist*. Pulau Pinang, Malaysia: Universiti Sains Malaysia press.

Alkarkhi, A. F. M., Saifullah, B. R., Yeoh, S. Y., & Azhar, M. E. (2011). Comparing physicochemical properties of banana pulp and peel flours prepared from green and ripe fruits. *Food Chemistry, 129*, 312–318.

Allan, G. B. (2007). *Elemantary statistics: A step by step approach*. McGraw-Hill.

Ben, B. (2010). *The many flavors of apply* [Online]. Available: <http://ms.mcmaster.ca/∼bolker/classes/m2e03/labs/lab1X.html> Accessed 26.10.17.

Bryan, F. J. M. (1991). *Multivariate statistical methods : A primer*. Great Britain: Chapman & Hall.

Chi, Y. *R Tutorial, An R introduction to Statistics* [Online]. Available: <http://www.r-tutor.com/elementary-statistics/numerical-measures/mean> Accessed 15.04.17.

Daniel, H. (2013). *Blog Archives, High Resolution Figures in R* [Online]. Available: <https://www.r-bloggers.com/author/daniel-hocking/> Accessed 15.07.17.

DATACAMP. *Tutorial on the R Apply Family, The Apply Functions As Alternatives To Loops* [Online]. DataCamp. Available: <https://www.datacamp.com/community/tutorials/r-tutorial-apply-family#gs.rqdg = Dw> Accessed 08.09.17.

Hadley, W. *Advanced R, Functionals* [Online]. Available: <http://adv-r.had.co.nz/Functionals.html> Accessed 18.10.17.

Johnson, R. A., & Wichern, D. W. (2002). *Applied multivariate statistical analysis*. New Jersey: Prentice Hall.

Kabacoff, R. I. (2014). *Quick R, accessing the power of R* [Online]. Available: <http://www.statmethods.net/graphs/scatterplot.html> Accessed 07.08.16.

Kelly, B. (2015). *R tutorial, Basic Operations and Numerical Descriptions* [Online]. Available: <http://www.cyclismo.org/tutorial/R/basicOps.html> Accessed 07.07.16.

Mark, G. *Using R for statistical analyses - Introduction* [Online]. Available: <http://www.gardenersown.co.uk/education/lectures/r/> Accessed 05.04.16.

Rencher, A. C. (2002). *Methods of multivariate analysis*. New York: J. Wiley.

William, B. K. (2016). *Tutorials* [Online]. Available: <http://ww2.coastal.edu/kingw/statistics/R-tutorials/index.html> Accessed 05.08.16.

Chapter 5

Hypothesis Testing

Learning outcomes

At the end of this chapter, you should be able

- *To describe hypothesis testing.*
- *To formulate null and alternative hypotheses.*
- *To conduct hypothesis testing with respect to a mean value for one sample (the univariate case).*
- *To conduct hypothesis testing with respect to a mean vector for one sample (the multivariate case).*
- *To perform hypothesis testing to compare two population means (with one variable).*
- *To perform hypothesis testing to compare two multivariate population means.*
- *To understand the difference between univariate and multivariate tests.*
- *To use built-in functions and R commands to perform hypothesis testing.*
- *To understand and interpret the results to make a decision.*
- *To report useful conclusions regarding hypothesis testing.*

5.1 WHAT IS HYPOTHESIS TESTING?

An important part of any analysis is to make inferences about the problem under study to make conclusions related to a population. To make a decision about any matter, researchers must have enough information about a topic to decide correctly. The decision is usually made by information collected from a representative sample of the population. Hypothesis testing is one of the most valuable tools that guide researchers to make the correct decision about any study and in any field. Hypothesis testing is easier to perform on univariate data than multivariate data because multivariate studies require methods that consider the correlation between different variables. Hypothesis testing concerning univariate distributions and the corresponding tests for multivariate distributions will be given in this chapter.

5.2 HYPOTHESIS TESTING IN R

R provides a remarkable variety of functions to cover all of the common statistical analysis techniques in the basic R installation. Furthermore, many additional packages have been added to provide additional functions beyond the standard packages (base installation). The functions and commands used to perform hypothesis testing in R will be presented with their structure and an explanation of each function.

1. The function `t.test()` is used to compute a variety of t-test statistics for one sample and for two samples.

```
t.test (data frame)          # H_0:mu = 0
t.test (data frame,mu = value)        # H_0:mu = true value of the mean
```

```
t.test (data frame, mu = value, conf.level = alpha, alternative = "less", mu = value)
```

The alternative hypothesis can be set as `alternative = "greater"` instead of `"less"`.

Easy Statistics for Food Science with R. DOI: https://doi.org/10.1016/B978-0-12-814262-2.00005-4

2. The function `t.test()` is also used to compute a t-test statistic for two independent samples, where y_1 and y_2 are numeric variables.

```
t.test(y1,y2)
```

We can include the option `alternative = "less"` or `"greater"`, `mu = 0`, and equal variances can be added (`vari.equal = TRUE`).

3. The function `t.test()` is also used to calculate a t-test statistic for two independent samples, where y is numeric and X is a binary factor.

```
t.test(y~x)
```

4. The function `t.test()` is used to compute a t-test statistic for two paired samples (grouped samples), where y_1 and y_2 are numeric.

```
t.test(y1, y2, paired = TRUE)
```

5. The function `sqrt()` is used to calculate the square root.

```
sqrt(data frame)
```

6. The function `t()` is used to produce the transpose of a matrix.

```
t(data frame)
```

7. The function `qt()` is called the quantile function. It is used to calculate the quantiles and critical values of a t distribution.

```
qt(data frame)
```

8. The function `dnorm()` is used to compute the height (probability) of the normal curve (the density function, (pdf)).

```
dnorm(x, mean = 0, sd = 1, log = FALSE)
```

For a standard normal distribution, the mean and standard deviation are set to 0 and 1, respectively. Other values for the mean and standard distribution can be given to calculate the probability. `x` indicates that a variable is being evaluated.

9. The function `pnorm()` is used to calculate the cumulative probability (the area below a given value of "X" (cdf), i.e., to the left of the q value).

```
pnorm(q, mean = 0, sd = 1, lower.tail = TRUE, log.p = FALSE)
```

The option `lower.tail = FALSE` means calculate the area to the right of q, whereas the option `TRUE` means calculate the area to the left of q.

10. The `qnorm()` function is the inverse of `pnorm()`. This function is used to obtain quantiles or `"critical values"`; `qnorm()` deals by default with areas below the given boundary value.

```
qnorm(p, mean = 0, sd = 1, lower.tail = TRUE, log.p = FALSE)
```

Where p represents the area under the standard normal curve.

11. The function `solve()` is used to compute the inverse matrix.

```
solve(data frame)
```

5.3 GENERAL PROCEDURE FOR HYPOTHESIS TESTING

Before discussing the general procedure for hypothesis testing, we must give some important definitions and terms that are related to hypothesis testing.

5.3.1 Definitions

A statistical hypothesis is defined as any claim about a population parameter; this claim may or may not be true.

Two types of statistical hypotheses can be recognized in hypothesis testing: The first type is called the *null hypothesis*, and the second type is called the *alternative hypothesis*.

The *null hypothesis* is denoted by H_0, and it is defined as a statistical hypothesis stating that there is no real difference (significant difference) between a population parameter (such as the mean, variance, or proportion) and the claimed value. In other words, it states that the difference between a population parameter and the claimed (observed) value is due to chance (the real difference is equal to zero).

The *alternative hypothesis* is denoted by H_1 or H_a, and it is defined as a statistical hypothesis stating that there is a real difference (significant difference) between a population parameter and the claimed value. In other words, it states that the difference between a population parameter and the claimed (observed) value is true and real. This event would imply that the parameter has a value that differs from the null hypothesis. The alternative hypothesis is considered as the inverse of the null hypothesis.

Example 5.1: State the null and alternative hypotheses- State the null and alternative hypotheses for three different claims regarding the pH value in a green banana peel.

a. The average value of the pH in a green banana peel is 5.
b. The average value of the pH in a green banana peel is greater than 5.
c. The average value of the pH in a green banana peel is less than 5.

Three different situations will be considered for formulating hypotheses.

a. Based on the claim that the average value of the pH in a green banana peel is exactly 5 (the mean is equal to 5), the null hypothesis is:

$$H_0:\mu = 5(\text{claim})$$

The alternative hypothesis, which is the opposite of the null hypothesis, will be either more or less than 5, which means that the average value is not equal to 5.

$$H_1:\mu \neq 5$$

It can be seen that the two hypotheses are formulated in such a way that they are mutually exclusive. That is, if one is true, the other must be false, and vice versa.

b. The average value of the pH in a green banana peel is greater than 5.

Here, the claim is "more than 5 and does not equal 5." As long as the claim does not contain the equality sign, then the claim will be under the alternative hypothesis (without an equality sign), and if the result is not greater than 5, then the result will be either equal to 5 or less than 5. This direction will be under the null hypothesis.

$$H_0: \mu \leq 5 \quad vs \quad H_1: \mu > 5(Claim)$$

c. The average value of the pH is less than 5.

Here, the claim is "less than 5" and does not contain a hypothesis of equality. Observe that the claim will compose the alternative hypothesis, hence the opposite of the claim is "greater than or equal to 5," which will be under the null hypothesis.

$$H_0: \mu \geq 5 \quad vs \quad H_1: \mu < 5(Claim)$$

We have given three different situations with statistical hypotheses. One type of these hypotheses must be used to test any problem.

The three situations of the statistical hypothesis are summarized below.

Two-tailed test	One-tailed test	
	Right-tailed test	Left-tailed test
$H_0: \mu = C$	$H_0: \mu \leq C$	$H_0: \mu \geq C$
$H_1: \mu \neq C$	$H_1: \mu > C$	$H_1: \mu < C$

where C is a given value.

Note
- For the null hypothesis, H_0, we always use the equality sign.
- For alternative hypothesis, H_1, we always use either $<, >$, or $\neq$.

5.3.2 Definitions

The *test statistic* is defined as a value computed from the sample values and used in hypothesis testing to make a decision regarding whether to support or reject the null hypothesis.

The *critical region* (rejection region) is defined as the set of the values of the test statistic that lead to the rejection of the null hypothesis.

The *noncritical region* (nonrejection region) is defined as the set of values of the test statistic that indicate that the null hypothesis should not be rejected.

The *critical value* is defined as a value that separates the rejection region (where the null hypothesis should be rejected) from the nonrejection region.

The *significance level* denoted by α is defined as the probability that the test statistic will fall in the rejection region (critical region) when the null hypothesis is true (no difference). The researchers usually choose α to be 0.05, 0.01, or 0.10.

The *p-value* is used in hypothesis testing to help make a decision to support or reject the null hypothesis. The *p-value* is defined as the probability of obtaining a more extreme value of the test statistic than the observed value when the null hypothesis is true. The p-value is a number between 0 and 1. Whereas a small p-value leads one to reject the null hypothesis, a large p-value will not permit one to reject the null hypothesis.

Example 5.2: Identifying critical and noncritical regions—Identify the rejection and nonrejection regions for each of the alternative hypotheses. Use a significance level $\alpha = 0.05$. Assume that the distribution is normal.

$$a. H_1: \mu \neq 7$$

$$b. H_1: \mu > 7$$

$$c. H_1: \mu < 7$$

a. For the first hypothesis $H_1: \mu \neq 7$, the critical region (rejection region) is in both tails of the normal distribution, where each tail contains an area of $(0.05/2 = 0.025)$. The Z value is ± 1.96, as shown in Fig. 5.1A (see the appendix for the corresponding R code).

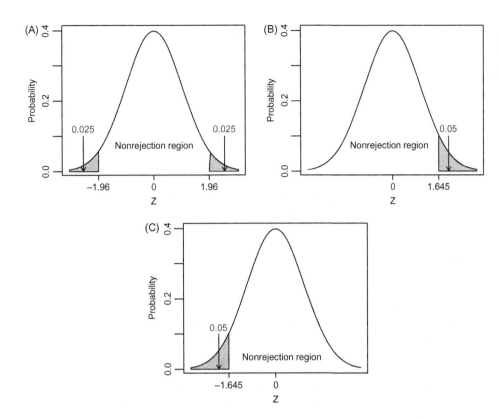

FIGURE 5.1 The rejection and nonrejection regions (A) The rejection and nonrejection region for $H_1 : \mu \neq 7$; (B) The rejection and nonrejection regions for $H_1 : \mu > 7$; (C) The rejection and nonrejection regions for $H_1 : \mu < 7$.

b. For the second hypothesis $H_1 : \mu > 7$, the critical region is in the right tail of the normal distribution and has a right-tail area of 0.05. The Z value is 1.645, as shown in Fig. 5.1B (see the appendix for the R code).

c. For the third hypothesis $H_1 : \mu < 7$, the critical region is in the left tail of the normal distribution and has a left-tail area of 0.05. The Z value is -1.645, as shown in Fig. 5.1C (see the appendix for the R code).

5.4 HYPOTHESIS TESTING ABOUT A MEAN VALUE

Hypothesis testing for a claim about the mean value of the data in both cases of one variable and several variables will be presented in this section.

5.4.1 Inference About a Mean Value for One Sample (Univariate)

In a case with only one variable measured for each experimental unit, hypothesis testing for the mean value consists of two cases, based on whether the variance is known or unknown. The two cases are given below.

a. When σ is known

Hypothesis testing about a mean value of a variable when the variance is known will only be studied for two-sided (two-tailed) because the alternative hypothesis cannot be generalized to multivariable tests for other cases. The researchers are interested in testing the hypothesis in (5.1).

$$H_0 : \mu = \mu_0 \quad \text{vs} \quad .H_1 : \mu \neq \mu_0 \tag{5.1}$$

The null hypothesis (H_0) states that the mean value of the variable is equal to a given value (μ_0), while the alternative hypothesis (H_1) states that the mean value differs from the given value (μ_0), which means that the mean value is either higher or lower than μ_0.

Assume that a random sample of size n is normally distributed. The test statistic used to test the equality of a mean to a given value is called the Z-test and is defined in (5.2).

$$Z = \frac{\overline{Y} - \mu_0}{\sigma/\sqrt{n}}$$

(5.2)

where

$\overline{Y}$ is the mean value,
μ_0 is a given value,
σ is the standard deviation, and
n is the sample size (the total number of observation).

Making a decision depends on the value of Z-test statistic and the critical value (theoretical value) of Z. If $|Z| \geq Z_{\alpha/2}$, the null hypothesis (H_0) is rejected, which indicates that the mean value differs from the given value.

Example 5.3: The pH value of banana flour—A nutritionist claims that the mean value of the pH of banana flour is six. Thirty-six samples were selected to test the nutritionist's claim. The sample data showed that the average pH is 5.11. The standard deviation of the population is 0.44. Is there sufficient evidence to support the claim? Use $\alpha = 0.05$. Assume that the population is normally distributed.

The null and alternative hypotheses as defined in (5.1) are:

$$H_0{:}\mu = 6 \quad vs \quad H_1{:}\mu \neq 6$$

The Z-test in (5.2) can be used to test the hypothesis about the mean pH value obtained from banana flour. The Z-test is not included in the default R packages. Thus built-in functions and R commands should be given and used to perform a Z-test. To illustrate the calculation of the Z-test in R, the mean, hypothesized mean value, standard deviation, and sample size should be provided to apply the formula given in (5.2). The R commands with the results for testing the pH value in banana flour using the Z-test are given below.

```
mean = 5.11        # sample mean
mu = 6             # hypothesized value
s = 0.44           # population standard deviation
n = 36             # sample size
z = (mean - mu)/( s / sqrt(n))
z                  # print z-test statistic
alpha = 0.05
z1 = round (qnorm (1 - alpha/2) , digits = 2) # critical value
c.vs = c(-z1, z1)        # critical value for two sided
c.vs               #Print critical values
```

The first four lines of commands provide the necessary information to apply the Z formula in (5.2). The results produced by R commands for a two-sided Z-test with critical values for the pH value of banana flour are given below.

```
> mean = 5.11                # sample mean
> mu = 6                     # hypothesized value
> s = 0.44                   # population standard deviation
> n = 36                     # sample size
> z = (mean - mu)/( s / sqrt(n))
> z                          # print z-test statistic
[1] -12.13636
> alpha = 0.05
> z1 = round (qnorm (1 - alpha/2) , digits = 2) # critical value
> c.vs = c(-z1, z1)          # critical value for two sided
> c.vs                       # Print critical values
[1] -1.96 1.96
```

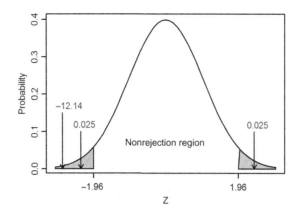

FIGURE 5.2 The rejection and nonrejection regions for the pH value in banana flour.

The first four lines of the R output represent the input information including the mean, hypothesized value, standard deviation, and number of observations. The fifth line shows the Z-test formula (5.2) and prints the result: $-12.13636 = -12.14$. The significance level alpha = 0.05 was chosen to calculate the critical values `z1` using the function `qnorm()`. The critical values for a two-sided Z-test were rounded to two decimals and are -1.96 and 1.96. To make a decision on the nutritionist's claim, the absolute calculated value of the test statistic Z should be compared with the critical (theoretical) value of 1.96, or one can identify the rejection and non-rejection regions on the graph as shown in Fig. 5.2. By numerical comparison, the null hypothesis is rejected because $|-12.14| > 1.96$. Alternately, the results of the Z-test for the pH value and the critical values are presented in Fig. 5.2 (see the appendix for the R code); clearly, the value -12.14 falls in the rejection region. Thus the decision is to reject the null hypothesis. The required graph with the summary statistics for pH is shown in Fig. 5.2.

In conclusion, there is sufficient evidence to reject the nutritionist's claim that the average pH value in banana flour is six and believe that the average mean value of the pH differs from six.

Example 5.4: Fat content in deep-fried snacks—A researcher wants to investigate the claim that the mean value of the fat content (%) in tapioca deep-fried chips is 19. Forty-two samples were collected and tested for their fat content. The results revealed that the average fat content was 27. A previous study showed that the variance of the population is 5. Is there sufficient evidence to support the claim? Use $\alpha = 0.01$. Assume that the population is normally distributed.

The claim states that the average fat content in tapioca deep-fried chips is 19. Thus the null and alternative hypotheses are as follows:

$$H_0: \mu = 19 \text{(Claim)} \quad \text{vs} \quad H_1: \mu \neq 19$$

Built-in R functions were used to compute the Z-test statistic using the formula given in (5.2) for the fat content in tapioca deep-fried chips. The commands to test the hypothesis regarding the fat content in tapioca chips are shown below.

```
mean = 27                # sample mean
mu = 19                  # hypothesized value
s = 5                    # standard deviation
n = 42                   # sample size
z = (mean - mu)/(s / sqrt(n))
z                        # print z-test statistic
alpha = 0.01
z1 = round (qnorm (1 - alpha/2) , digits = 2) # critical value
c.vs = c(-z1, z1)        # critical value for two sided
c.vs                     #Print critical values
```

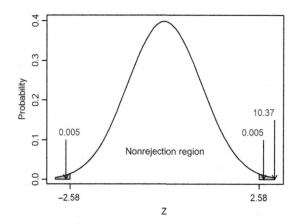

FIGURE 5.3   The rejection and nonrejection regions for the fat content in deep-fried food.

The results of the R commands for testing the claim that the fat contentin deep-fried food is 19 are given below.

```
> mean = 27               # sample mean
> mu = 19                 # hypothesized value
> s = 5                   # standard deviation
> n = 42                  # sample size
> z = (mean - mu)/(s / sqrt(n))
> z                       # print z-test statistic
[1] 10.36919
> alpha = 0.01
> z1 = round (qnorm (1 - alpha/2) , digits = 2) #critical value
> c.vs = c(-z1, z1)       # critical value for two sided
> c.vs                    #Print critical values
[1] -2.58 2.58
```

A decision should be made by comparing the absolute calculated Z-test statistic $10.36919 = 10.37$ with the critical value of 2.58. This comparison will lead to the rejection of the null hypothesis because $|10.37| > 2.58$. Thus there is sufficient evidence to believe that on average, the percentage of fat content in tapioca deep fried food is not 19. As in Example 5.1, a normal curve can be used to make a decision. R code was used to produce Fig. 5.3 (see the appendix for the R code) with the value of the Z-test statistic, critical values, and rejection and nonrejection regions. It can be seen that the value 10.37 falls in the rejection region.

b.   When σ is unknown

The Z-test was used to study the first case in hypothesis testing regarding the mean value. The Z-test is not valid when the variance of a population is unknown. A new test should be used in the case of unknown variance. Assume that a random sample $Y_1, Y_2, \ldots, Y_n$ is normally distributed. The test statistic used to test the equality of a mean to a given value when σ unknown is called the t-test, as given in (5.3).

$$t = \frac{\overline{Y} - \mu_0}{S/\sqrt{n}}, \text{with } n - 1 \text{degrees of freedom} \tag{5.3}$$

where

$\overline{Y}$ is the mean value,
μ_0 is a given value,
n is the sample size (total number of observations),
S is the sample's standard deviation.

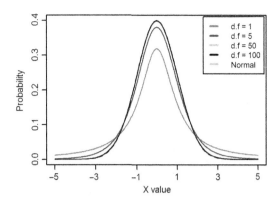

FIGURE 5.4   T distribution graph at different degrees of freedom.

The null hypothesis (H_0) is rejected if the calculated t value in (5.3) is greater than or equal to the tabulated (theoretical) value, $|t| \geq t_{\alpha/2, n-1}$, where $n - 1$ is the degree of freedom.

Note
- $t_{\alpha/2, n-1}$ is the critical value obtained from the t table.
- The null and alternative hypotheses are the same as those defined in (5.1).
- The t-test is used when the sample size is small, $n < 30$.
- The graph for t distribution for different degrees of freedom can be produced using R commands. The graph for t distribution is given in Fig. 5.4 (see the appendix for R codes).

Example 5.5: Zinc concentration in cockles—A researcher wants to verify the claim that the average concentration of zinc (Zn) in cockles is 0.15 (mg/L). Ten sampling points were selected, and the concentrations of Zn in cockles were recorded. The results showed that the average is 0.23 and that the standard deviation is 0.02. Use $\alpha = 0.05$ to test the researcher's claim, where we assume that the population is normally distributed.

The null hypothesis states that the concentration of Zn in cockles is 0.15 (mg/L), and the alternative hypothesis is that the average is more or less than 0.15 (mg/L). Thus the alternative hypothesis states that the average is not equal to 0.15 (mg/L), as given below.

$$H_0: \mu = 0.15 (\text{Claim}) \quad \text{vs} \quad H_1: \mu \neq 0.15$$

The function `t.test()` in R cannot be used for this example because the function `t.test()` is used with the raw data (the actual values of the study), but this example provided a summary of the concentrations of Zn in cockles. (This case is not included in the R default packages.) Thus built-in functions should be used, and R commands should be applied to perform the test using the formula given in (5.3). These commands are similar to the steps given for the Z-test. The function `qt()` was used to calculate the critical values for the t distribution. The results of using R commands for a two-sided t-test are given below.

```
#Two-sided t-test
mean = 0.23              # sample mean
mu = 0.15                # hypothesized value
s = 0.02                 # standard deviation
n = 10                   # sample size
t = (mean - mu)/(s / sqrt(n))
t                        # print t-test statistic
alpha = 0.05
c.v = qt (1 - alpha/2, df = n-1)       # critical value
t1 = round (c.v, digits = 2)
c.vs = c(-t1, t1)        # critical value for two sided
c.vs                     #Pring critical values
```

The output of the R commands used to perform a two-sided t-test for the concentration of Zn in cockles is given below.

```
> #Two-sided t-test
> mean = 0.23                    # sample mean
> mu = 0.15                      # hypothesized value
> s = 0.02                       # standard deviation
> n = 10                         # sample size
> t = (mean - mu)/(s / sqrt(n))
> t                              # print t-test statistic
[1] 12.64911
> alpha = 0.05
> c.v = qt (1 - alpha/2, df = n-1)      # critical value
> t1 = round (c.v, digits = 2)
> c.vs = c(-t1, t1)              # critical value for two sided
> c.vs                           #Pring critical values
[1] -2.26 2.26
```

The output is similar to the Z-test output in Example 5.1, and the only difference is in the entry data. The result showed that the null hypothesis should be rejected because the calculated t value 12.64911 = 12.65 is greater than the critical value of 2.26. This result indicates that the average Zn concentration in cockles is not 0.15 mg/L.

This decision can be achieved by producing a graph of the rejection and non-rejection regions. The rejection and nonrejection regions for the zinc concentration in cockles are given in Fig. 5.5.

Because the test statistic value of 12.65 is in the rejection (critical) region, the null hypothesis is rejected, as shown in Fig. 5.5. This rejection indicates that the concentration of Zn in cockles differs from the given value of 0.15 mg/L.

There is sufficient evidence to reject the researcher's claim (null hypothesis) and conclude that the zinc concentration in cockles is not equal to 0.15 (mg/L).

Example 5.6: Total phenolic content value in dates—A researcher in food science claimed that the average total phenolic content (TPC) value in dates is 5 mg gallic acid equivalents (GAE)/100 g. Seven types of dates were selected, and three samples were tested for the TPC from each type using Folin−Ciocalteau. The results of the analysis showed that the average TPC is 3.2 (GAE)/100 g and the standard deviation is 1.13 (GAE)/100 g. Assuming that the population is normally distributed, use $\alpha = 0.10$ to test the researcher's claim.

First, state the hypotheses and verify the claim:

$$H_0: \mu = 5(\text{claim}) \quad \text{vs} \quad H_1: \mu \neq 5$$

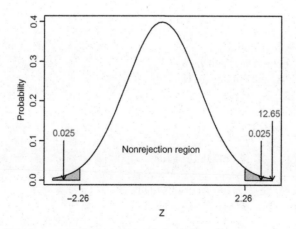

FIGURE 5.5 The rejection and nonrejection regions for the Zn concentration in cockles.

The calculated t-test value can easily be achieved using R commands similar to Example 5.5 after accounting for the information for the TPC value in dates. A t-test was performed using the R statistical software to test the hypothesis on the TPC in dates. The results of using R commands to test the hypothesis on the total phenolic content with the t-test are given below.

```
> mean = 3.2             # sample mean
> mu = 5                 # hypothesized value
> s = 1.13               # standard deviation
> n = 21                 # sample size
> t = (mean - mu)/( s/ sqrt(n))
> t                      # print t-test statistic
[1] -7.299678
> alpha = 0.10
> c.v = qt (1 - alpha/2, df =n-1)      # critical value
> t1 = round (c.v, digits = 2)
> c.vs = c(-t1, t1)      # critical value for two sided
> c.vs                   #Pring critical values
[1] -1.72 1.72
```

The null hypothesis is rejected because the absolute value for the test statistic value |7.299678| = 7.30, which is greater than the critical value of 1.7 and is in the rejection (critical) region (Fig. 5.6). There is sufficient evidence not to support the claim that the TPC value in dates is equal to 5 (GAE)/100 g.

Note
- If raw data are used, then the function t.test () is used to carry out the test.

```
t.test (data frame)
```

- If the null hypothesis is different from zero, the following command is used.

```
t.test (y, mu = value) # Ho: mu = hypothesized value
```

```
t.test (data frame, mu = value, alternative = "two.sided")
```

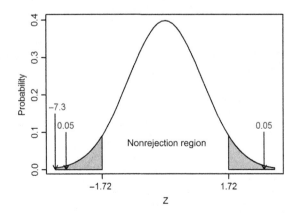

FIGURE 5.6 The rejection and nonrejection regions for the total phenolic content value in dates.

TABLE 5.1 Glucose Values in Commercial Orange Juice

23.59	23.78	35.98	36.31	20.20	20.40	22.66	22.25	20.72	20.62
23.07	23.69	5.37	5.19	5.89	6.06	15.72	15.82	17.07	17.00

Note
- Setting the alternative to `"two.sided"` was unnecessary, as that setting is the default. The alternative hypothesis can be set as `"less"` or `"greater"`. For example:

```
t.test (data frame, mu = value, alternative = "less")
```

Example 5.7: The glucose value in commercial orange juice—The value of glucose (mg/mL) in commercial orange juice was measured in 20 samples including different brands. The data are given in Table 5.1.

Test the following hypotheses regarding the average quantity of glucose in commercial orange juice.

$$a.\ H_0 : \mu = 0 \ \text{vs} \ H_1 : \mu \neq 0$$

$$b.\ H_0 : \mu = 0.10 \ \text{vs} \ H_1 : \mu \neq 0.10$$

$$c.\ H_0 : \mu \leq 0.10 \ \text{vs} \ H_1 : \mu > 0.10$$

$$d.\ H_0 : \mu \geq 0.10 \ \text{vs} \ H_1 : \mu < 0.10$$

The function `t.test()` will be used to test the above hypotheses for the value of glucose in commercial orange juice.

a. When the null hypothesis is equal to zero:

$$H_0 : \mu = 0 \ \text{vs} \ H_1 : \mu \neq 0$$

The data for the glucose values in commercial orange juice had been stored as a .CSV (`Example5_7`) file. Thus the file should be imported first. Then we can run the commands to test the hypothesis $H_0 : \mu = 0$.

```
t.test (Example5_7)
```

The results of using the function `t.test()` to carry out a t-test to determine whether the glucose value differs from 0 are shown below.

```
> t.test (Example5_7)
        One Sample t-test
data: Example5_7
t = 9.77, df = 19, p-value = 7.646e-09
alternative hypothesis: true mean is not equal to 0
95 percent confidence interval:
14.98425 23.15475
sample estimates:
mean of x
19.0695
```

The first row of the output is a call to perform a t-test for the data that are stored in the file Example5_7. Then we have a statement to inform us that this test is a one-sample test. The information on the t-test statistic (t), degrees of freedom (df), and p-value are presented below "data: Example5_7" in the output. The alternative hypothesis of "true mean is not equal to 0" tells us that the hypothesis is a two-sided test. The last two results are the "95 percent confidence interval" and the mean of the variable (mean of x).

The decision is to reject the null hypothesis, which states that the average value of glucose is 0 since the p-value is very small at 7.646e-09, which is less than $\alpha = 0.05$.

b. The second case is when the null hypothesis is equal to a specified value.

$$H_0: \mu = 0.10 \ \text{vs} \ H_1: \mu \neq 0.10$$

The function t.test() is used to perform the test for this hypothesis.

```
t.test (Example5_7, mu = 0.10)
```

The results of applying the function t.test() are given below.

```
> t.test (Example5_7, mu = 0.10)
      One Sample t-test
data: Example5_7
t = 9.7188, df = 19, p-value = 8.316e-09
alternative hypothesis: true mean is not equal to 0.1
95 percent confidence interval:
14.98425 23.15475
sample estimates:
mean of x
19.069
```

The interpretation of the results is similar to the first case when the null hypothesis is equal to zero.

The null hypothesis, which states that the average value of glucose in commercial orange juice is 0.10, is rejected since the p-value of 8.316e-09 is very small and much less than $\alpha = 0.05$.

c. The third case is when the null hypothesis is less than or equal to 0.10.

$$H_0: \mu \leq 0.10 \ \text{vs} \ H_1: \mu > 0.10$$

The function t.test() is also used to test the hypothesis of "less than or equal." The procedure for testing this hypothesis is similar to the earlier cases a and b, and the only difference is to set the alternative hypothesis to "greater".

```
t.test (Example5_7, mu = 0.10, alternative = "greater")
> t.test (Example5_7, mu = 0.10, alternative = "greater")
      One Sample t-test
data: Example5_7
t = 9.7188, df = 19, p-value = 4.158e-09
alternative hypothesis: true mean is greater than 0.1
95 percent confidence interval:
15.6945      Inf
sample estimates:
mean of x
19.0695
```

Notice that the output is similar in content to the first two cases.

The decision is to reject the null hypothesis, which states that the average value of glucose is less than or equal to 0.10.

d. The fourth case is when the null hypothesis is greater than or equal to 0.10.

$$H_0: \mu \geq 0.10 \text{ vs } H_1: \mu < 0.10$$

The structure of the R commands is similar to the third case. In fact, the only difference is to set the alternative hypothesis differently.

```
t.test (Example5_7, mu = 0.10, alternative = "less")
```

The results of applying the function t.test()are similar to the other t-test outputs.

```
> t.test (Example5_7, mu = 0.10, alternative = "less")
    One Sample t-test
data: Example5_7
t = 9.7188, df = 19, p-value = 1
alternative hypothesis: true mean is less than 0.1
95 percent confidence interval:
        -Inf 22.4445
sample estimates:
mean of x
19.0695
```

Observe that the same function is used in R to carry out a t-test for all of the cases, and the only difference is to change the alternative option.

The decision is not to reject the null hypothesis that the average value of glucose is greater than or equal to 0.10, as the p-value is greater than 0.05.

5.4.2 Inference About a Mean Vector for One Sample (Multivariate)

Hypothesis testing regarding one variable measured for each sampling unit has been described. We will make inferences about a mean vector when there are several variables (k) measured for each sampling unit. In this case, we are interested in testing the hypotheses in (5.4).

$$H_0: \boldsymbol{\mu} = \boldsymbol{\mu_0} \text{ vs } H_1: \boldsymbol{\mu} \neq \boldsymbol{\mu_0}$$

$$H_0: \begin{bmatrix} \mu_1 \\ \mu_2 \\ \vdots \\ \mu_k \end{bmatrix} = \begin{bmatrix} \mu_{01} \\ \mu_{02} \\ \vdots \\ \mu_{0k} \end{bmatrix} \text{ vs } H_1: \begin{bmatrix} \mu_1 \\ \mu_2 \\ \vdots \\ \mu_k \end{bmatrix} \neq \begin{bmatrix} \mu_{01} \\ \mu_{02} \\ \vdots \\ \mu_{0k} \end{bmatrix} \qquad (5.4)$$

The null hypothesis in (5.4) states that the mean value of each variable equals a given value (specified value), while the alternative hypothesis states that at least one mean value does not equal its given value.

The test statistic used for testing a hypothesis in the case of several variables is called Hotelling's test, which is an extension of the t-test for one variable. Assume that there are k variables (normally distributed) being considered, with sample size n. The Hotelling's test (T^2) for testing k variables measured for each sampling unit is given in (5.5).

$$T^2 = n(\overline{Y} - \boldsymbol{\mu_0})'S^{-1}(\overline{Y} - \boldsymbol{\mu_0}) \qquad (5.5)$$

where $\overline{Y}$ is a vector of the means and $\boldsymbol{\mu_0}$ is a vector of the given values.

The null hypothesis is rejected if $T^2 > T^2_{\alpha,k,n-1}$, where $T^2_{\alpha,k,n-1}$ is the tabulated value (critical value).

Example 5.8: Physical and chemical properties of noodles—The physical and chemical properties of noodles made from ripe banana pulp (Cavendish) were studied. Three parameters were measured from each sample: the pH, color (L^*), and tensile strength (TS) (kPa). Six replicates were used, and the results were recorded. The data are given in Table 5.2.

TABLE 5.2 The Results of the Physical and Chemical Properties for Noodles

pH	L*	Ts
7.62	49.95	77.65
7.51	49.95	66.66
7.63	43.69	78.63
7.44	44.45	78.71
7.64	50.74	72.37
7.58	49.61	77.62

Is there sufficient evidence to say that $\mu_1 = 7, \mu_2 = 49, \mu_3 = 73$ for pH, L*, and TS, respectively? Use $\alpha = 0.05$. First, the null and alternative hypotheses should be stated as in (5.4).

$$H_0: \begin{bmatrix} \mu_1 \\ \mu_2 \\ \mu_3 \end{bmatrix} = \begin{bmatrix} 7 \\ 49 \\ 73 \end{bmatrix} \text{ vs } H_1: \begin{bmatrix} \mu_1 \\ \mu_2 \\ \mu_3 \end{bmatrix} \neq \begin{bmatrix} 7 \\ 49 \\ 73 \end{bmatrix}$$

Hotelling's test (T^2) is used for this case since there is more than one variable (parameter). To obtain the test statistic value for T^2 as defined in (5.5), the variance-covariance matrix should be calculated first, along with the average value for each variable.

$$T^2 = n(\overline{Y} - \mu_0)' S^{-1}(\overline{Y} - \mu_0)$$

Hotelling's test T^2 is not included in the default packages. Thus we should give commands using built-in functions to perform Hotelling's test T^2. R commands were used to execute Hotelling's test, including the calculations of the averages, transpose, variance-covariance matrix, inverse matrix, and the value of Hotelling's test. These commands are given below. The data for the chemical and physical properties for noodles had been stored as a .CSV (Example5_8) file.

```
# calculate the mean vector (observed)
M <- round (sapply (Example5_8,mean), digits = 3)
M                              # print the mean vector
Mn <- c(7, 49, 73)                # Hypothesized mean vector
d = M - Mn                  # the difference between the mean and claim
d
dt = t(d)                      #transpose for the difference
dt
n <- length (Example5_8 $ pH)      #number of observations
n
co = round (cov(Example5_8), digits = 3)      # covariance matrix
co
IN = solve(co)                    # calculate the inverse
IN
HOT = dt%*%IN%*%d*n
HOT                            # Hotelling's value
```

The function `sapply()` is used when a similar task must be executed multiple times for all of the columns of an array or list. Detailed results for Hotelling's test were obtained by running R commands for the physical and chemical properties of the noodles.

```
> M <- round (sapply (Example5_8,mean), digits = 3)
> M                              # print the mean vector
   pH      L      Ts
 7.570 48.065 75.273
> Mn <- c(7, 49, 73)                     # Hypothesized mean vector
> d = M - Mn             # the difference between the mean and claim
> d
   pH      L      Ts
 0.570 -0.935 2.273
> dt = t(d)                      #transpose for the difference
> dt
    pH      L      Ts
[1,] 0.57 -0.935 2.273
> n <- length (Example5_8 $ pH)    #number of observations
> n
[1] 6
> co = round (cov(Example5_8), digits = 3)    # covariance matrix
> co
        pH        L      Ts
pH   0.006    0.078   0.042
L    0.078    9.772  -8.601
Ts   0.042   -8.601  23.370
> IN = solve(co)                 # calculate the inverse
> IN
            pH            L            Ts
pH   217.275793  -3.07363722  -1.52169179
L     -3.073637   0.19484599   0.07723419
Ts    -1.521692   0.07723419   0.07394961
> HOT = dt%*%IN%*%d*n
> HOT                            # Hotelling's value
    [,1]
[1,] 420.901
```

The average value for each variable, the difference between the hypothesized and calculated means, and the transposes of the difference vectors are calculated and presented first. Then we present the commands to calculate the number of observations (n) as well as the variance-covariance matrix and its inverse. The last result is for Hotelling's test (420.901).

The critical value is $T^2_{0.05,3,5} = 46.383$. The null hypothesis is rejected because the test statistic value (Hotelling's test) HOT = 420.901 is greater than the critical value of 46.383. This result indicates that on average, at least one parameter does not equal the hypothesized value. The values for pH, L*, and TS are $\mu_1 = 7, \mu_2 = 49, \mu_3 = 73$, respectively.

Example 5.9: Fat content and moisture in tapioca deep-fried snacks—The percentages of fat content and moisture (%) in tapioca deep-fried snacks were measured in 21 samples. The data for the fat content and moisture for 10 replicates are given in Table 5.3.

Is there sufficient evidence to say that $\mu_1 = 19, \mu_2 = 4$? Use $\alpha = 0.05$.

The null and alternative hypotheses for the mean value of the fat content and moisture in tapioca deep fried snacks are given in (5.4).

$$H_0 : \begin{bmatrix} \mu_1 \\ \mu_2 \end{bmatrix} = \begin{bmatrix} 19 \\ 4 \end{bmatrix} \text{ vs } H_1 : \begin{bmatrix} \mu_1 \\ \mu_2 \end{bmatrix} \neq \begin{bmatrix} 19 \\ 4 \end{bmatrix}$$

The command will be similar to the R commands used in Example 5.8 to compute the test statistic T^2, which is defined in (5.5). The input information to calculate the T^2 value will be for the fat content and moisture in tapioca deep fried snacks. The data had been stored as a .CSV (Example5_9) file. The results of running the requisite R commands are shown below.

TABLE 5.3 Fat Content and Moisture in Deep-Fried Snacks

Fat content	20.30	19.96	20.29	18.57	18.73	18.44	17.74	17.22	17.46	26.05	
Moisture	2.29	2.24	2.27	3.29	3.41	3.33	3.74	3.75	3.24	5.32	
Fat content	26.42	26.23	20.62	20.89	20.51	29.91	30.21	29.92	25.99	26.04	25.60
Moisture	5.47	5.39	2.79	2.62	2.57	4.57	4.60	4.55	4.55	4.47	4.47

```
> # calculate the mean vector (observed)
> M <- round(sapply(Example5_9,mean),digits = 3)
> M                             # print the mean vector
Fat.content      Moisture
   22.719          3.759
> Mn <- c(19, 4)                    # Hypothesized mean vector
> d = M - Mn                 # the difference between the mean and claim
> d
Fat.content      Moisture
   3.719          -0.241
> dt = t(d)                     #transpose for the difference
> dt
    Fat.content      Moisture
[1,]      3.719       -0.241
> n <- length (Example5_9 $ Fat.content)   #number of observations
> n
[1] 21
> co = round (cov(Example5_9), digits = 3)      # covariance matrix
> co
            Fat.content   Moisture
Fat.content    19.647       3.412
Moisture        3.412       1.139
> IN = solve(co)                  # calculate the inverse
> IN
            Fat.content   Moisture
Fat.content    0.1060898   -0.3178036
Moisture      -0.3178036    1.8299790
> HOT = dt%*%IN%*%d*n
> HOT                            # Hotelling's value
        [,1]
[1,] 45.00914
```

Here, the null hypothesis is rejected because the test statistic value (Hotelling's test) HOT = 45.009 is greater than the critical value $T^2_{0.05,2,20} = 7.415$, which indicates that, on average, the percentage of at least one parameter (the fat content and moisture) is not equal to the hypothesized values of $\mu_1 = 19, \mu_2 = 4$.

Note
- Sometimes the results of a univariate test do not match the results of a multivariate test. However, the multivariate results should be preferred.

5.5   COMPARING TWO POPULATION MEANS

The concept of hypothesis testing for one sample can be developed to test the differences between two sample means in both the univariate and multivariate cases. Sometimes researchers are interested in studying the differences between two groups of data about the same variable (variables) measured in each experimental unit to obtain a clear and comprehensive picture of the variable (variables) under study.

5.5.1   Comparing the Means of Two Populations (One Variable)

Consider two normally distributed populations. Two samples were selected, one from each population, with sample sizes n_1 and n_2. Assume that the two samples are independent and have equal variances; $\sigma_1^2 = \sigma_2^2 = \sigma^2$.

The researchers are interested in testing the hypotheses defined in (5.6).

$$H_0: \mu_1 = \mu_2 \ \text{ vs } \ H_1: \mu_1 \neq \mu_2 \tag{5.6}$$

or

$$H_0 : \mu_1 - \mu_2 = 0 \ \text{ vs } \ H_1 : \mu_1 - \mu_2 \neq 0$$

The null hypothesis states that the mean values of the two populations are the same, while the alternative hypothesis states that the mean values of the two populations are different.

The test statistic used to test the hypothesis of equality in (5.6) is called the t-test for two samples. The test statistic for testing two independent samples is defined in (5.7).

$$t = \frac{\overline{Y}_1 - \overline{Y}_2}{Sp\sqrt{\frac{1}{n_1} + \frac{1}{n_2}}} \tag{5.7}$$

with $d.f = n_1 + n_2 - 2$ where S_p is the pooled variance. The formula for computing the pooled variance is given in (5.8).

$$Sp = \sqrt{\frac{(n_1 - 1)S_1^2 + (n_2 - 1)S_2^2}{n_1 + n_2 - 2}} \tag{5.8}$$

The decision is to reject the null hypothesis if the absolute value of the calculated test statistic is greater than or equal to the tabulated value: $|t| \geq t_{\alpha/2, n_1+n_2-2}$, or if the test value falls in the rejection region.

Built-in functions can be used to calculate the t-test statistic value for two samples. The function `t.test()` is also used to perform a t-test for two samples. The structure of the function `t.test()` for testing two independent samples in R is as follows:

```
t.test(y1,y2)
```

where y_1 and y_2 are numeric variables.

Note
* The function `t.test()` can be used for two variables when one variable is binary.

```
t.test(y ~ x)
```

where y is numeric and X is binary.

Example 5.10: Total phenolic content (TPC) in honey and Zehdi dates—A researcher is interested in comparing the total phenolic content (TPC) (GAE)/100 g) in honey and Zehdi dates. Nine samples were selected from each category, and the total phenolic content valuewas tested. Assume that the two populations are normally distributed. The results are given in Table 5.4. Can we conclude that at $\alpha = 0.05$, the mean value of the TPC in the two types of dates is different?

TABLE 5.4 The Results for the TPC (GAE)/100 g) Values in Honey and Zehdi Dates

Honey	1.64	1.73	1.82	1.79	1.77	1.80	1.85	1.68	1.66
Zehdi	3.58	4.71	3.36	3.65	3.91	3.47	4.58	3.75	3.63

The objective of this example is to study the difference in the total phenolic content value in the two types of dates; thus the null and alternative hypotheses as defined in (5.6) are:

$$H_0: \mu_1 = \mu_2 \quad \text{vs} \quad H_1: \mu_1 \neq \mu_2$$

The null hypothesis states that there is no difference in the value of the TPC for the two types of dates, which means that the value of the TPC in honey dates (μ_1) is equal to the TPC in Zehdi dates (μ_2).

Computations of the test statistic as defined in (5.7) for two samples can be achieved using R commands that are similar to the commands used in Example 5.2. The data for the total phenolic content had been stored as a .CSV (Example5_10) file. The function t.test (y1, y2) is used to test the hypothesis regarding the total phenolic content based on the data for the two types of dates. In addition, the function qt() is used to calculate the t critical values.

```
t.test (Example5_10 $ Honey, Example5_10 $ Zehdi)
```

The data for the TPC in dates obtained from honey and Zehdi dates were analyzed using a t-test for two independent samples t.test(y1, y2).

```
> T <- t.test (Example5_10 $ Honey, Example5_10 $ Zehdi)
> T
    Welch Two Sample t-test
data: Example5_10$Honey and Example5_10$Zehdi
t = -13.007, df = 8.3918, p-value = 7.478e-07
alternative hypothesis: true difference in means is not equal to 0
95 percent confidence interval:
  -2.469302 -1.730698
sample estimates:
mean of x mean of y
1.748889 3.848889
```

The first part of the output shows that the test is about the "welch two sample t-test", which is followed by the name of the file and the two types of dates used in the analysis. The result of the t-test (t), df, and p-value are given immediately after the information on the test, and the data used in the analysis are provided.

The decision is to reject the null hypothesis since the p-value is very small: the p-value < 7.478e-07, which is less than 0.05. It can be concluded that the total phenolic content value is different in honey and Zehdi dates. This difference could be due to the type of date used since honey dates are considered soft dates (SD), whereas Zehdi dates are considered as semidry dates (SDD).

This result can be shown in a graph. R commands are used to produce a graph with the critical values, calculated values, rejection regions, and nonrejection regions, as illustrated in Fig. 5.7 (see the appendix for the R code).

Example 5.11: Concentrations of potassium in Cavendish and Dream bananas—A researcher wants to compare the concentrations of potassium (K) in Cavendish and Dream bananas (mg/100 mg). Eleven samples were selected from each type of banana and tested for their concentrations of potassium. Assume that the two populations are normally distributed. The data are given in Table 5.5. At $\alpha = 0.05$, is there sufficient evidence to say that the concentrations of potassium in Cavendish and Dream bananas are different?

The null and alternative hypotheses regarding the concentrations of potassium in bananas are defined in (5.6).

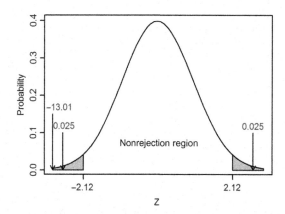

FIGURE 5.7 The rejection and nonrejection regions for the TPC content in honey and Zehdi dates.

TABLE 5.5 The Potassium (K) Concentrations in Cavendish and Dream Bananas (mg/100 g)

Cavendish	680.159	646.141	565.939	543.273	629.635	703.956
Dream	771.064	714.435	796.965	712.35	686.776	711.157
Cavendish	664.673	756.601	719.111	803.976	763.351	
Dream	690.398	756.973	791.091	589.326	474.598	

$$H_0: \mu_1 = \mu_2 \text{ vs } H_1: \mu_1 \neq \mu_2$$

Whereas the null hypothesis states that the concentration of potassium in Cavendish bananas (μ_1) is equal to the concentration of potassium in Dream bananas (μ_2), the alternative hypothesis states that the concentration of potassium is different in the two types of bananas. A t-test for two samples was carried out to test the equality of the means of potassium in Cavendish and Dream bananas. The t.test() function is used to perform the t-test for two independent samples and compute the test statistic as defined in (5.7). The data for the two types of banana had been stored as a .CSV (Example5_9) file.

```
t.test (Example5_11 $ Cavendish, Example5_11 $ Dream)
```

The results of applying a t-test for two independent samples using R commands are given below.

```
> T <- t.test (Example5_11 $ Cavendish, Example5_11 $ Dream)
> T
    Welch Two Sample t-test
data: Example5_11$Cavendish and Example5_11$Dream
t = -0.52751, df = 19.54, p-value = 0.6038
alternative hypothesis: true difference in means is not equal to 0
95 percent confidence interval:
   -98.44765 58.75347
sample estimates:
mean of x mean of y
   679.7105 699.5575
```

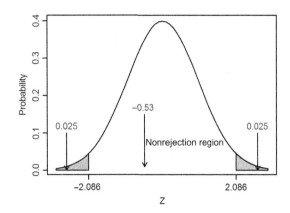

FIGURE 5.8 The rejection and nonrejection regions for the concentrations of potassium in bananas.

The results produced by applying R's built-in functions to test the concentrations of potassium in Cavendish and Dream bananas (t-test) showed that the null hypothesis is not rejected, as the `p-value` of `0.6038` is greater than 0.05. R commands are used to show the decision in a pictorial form, including the rejection and nonrejection regions with the critical values and calculated t value as illustrated in Fig. 5.8 (see the appendix for the R code).

It can be concluded that the variety of banana does not affect the concentration of potassium.

Note
- Sometimes the data are summarized as the mean, standard deviation, and number of observations. In this case, the function `t.test()` in R cannot be used to perform the t-test. However, R commands can still be used to calculate the t-test value and the critical values (see `Example5_11` in the appendix).

```
> M1 = 679.7105
> M2 = 699.5575
> n1 = 11
> n2 = 11
> df = n1 + n2 - 2
> M = M1 - M2
> s1 = 81.18496
> s2 = 94.76333
> sp = sqrt (((n1 - 1)*(s1^2) + (n2 - 1)*(s2^2))/df)
> sp
[1] 88.23573
> A = sqrt ((1/n1) + (1/n2))
> t = M/(sp*A)
> t
[1] -0.5275113
> alpha = 0.05
> c.v = qt(1 - alpha/2, df = n1 + n2 - 2)
> t1 = round (c.v, digits = 3)
> c.vs = c(-t1, t1)
> c.vs
[1] -2.086 2.086
```

The first part of the results includes the given information, such as the mean, standard deviation, and number of observations for each sample. The second part includes the calculations to carry out the test, and the last step is to calculate the critical values.

5.5.2 Comparing Two Multivariate Population Means

Assume that there are several variables measured in each experimental unit. The test statistic used to compare two multivariate means that considers the correlation between different variables is called the Hotelling's test (T^2).

Consider two samples that were selected from two multivariate normal populations, one sample from each population, with sample sizes n_1 and n_2. Assume that the two samples are independent and have equal variances, $\sum_1 = \sum_2 = \sum$. The hypotheses of interest, in this case, are defined in (5.9).

$$H_0: \boldsymbol{\mu_1} = \boldsymbol{\mu_2} \text{ vs } H_1: \boldsymbol{\mu_1} \neq \boldsymbol{\mu_2} \tag{5.9}$$

The hypotheses can be written as follows:

$$H_0: \begin{bmatrix} \mu_{11} \\ \mu_{21} \\ \vdots \\ \mu_{k1} \end{bmatrix} = \begin{bmatrix} \mu_{12} \\ \mu_{22} \\ \vdots \\ \mu_{k2} \end{bmatrix} \quad \text{vs} \quad H_1: \begin{bmatrix} \mu_{11} \\ \mu_{21} \\ \vdots \\ \mu_{k1} \end{bmatrix} \neq \begin{bmatrix} \mu_{12} \\ \mu_{22} \\ \vdots \\ \mu_{k2} \end{bmatrix}$$

The test statistic for comparing two multivariate means is given in (5.10).

$$T^2 = \frac{n_1 n_2}{n_1 + n_2} (\overline{Y}_1 - \overline{Y}_2)' Sp^{-1} (\overline{Y}_1 - \overline{Y}_2) \tag{5.10}$$

distributed as T_{α,k,n_1+n_2-2}, where Sp is the pooled variance and is calculated as defined in (5.11).

$$S_p = \frac{(n_1 - 1)S_1 + (n_2 - 1)S_2}{n_1 + n_2 - 2} \tag{5.11}$$

where S_1 and S_2 are the variance-covariance matrices.

The decision is to reject the null hypothesis if $T^2 \geq T^2_{\alpha,k,n_1+n_2-2}$.

Note
- It is necessary that $n_1 + n_2 - 2 > k$.

Example 5.12: Comparing heavy metals in cockles—A researcher seeks to compare the concentrations of arsenic (As) and six heavy metals, namely, chromium (Cr), cadmium (Cd), zinc (Zn), copper (Cu), lead (Pb), and mercury (Hg) (mg/L) in cockles obtained from two locations (the Juru and Jejawi Rives in the state of Penang Malaysia). Twenty samples were selected from each river and tested for the selected parameters. The data in Table 1.3 are reproduced in Table 5.6. At $\alpha = 0.05$, is there sufficient evidence that the concentrations of arsenic and heavy metals in cockles obtained from the Juru River are the same as those in the Jejawi River?

The null hypothesis states that there is no difference between cockles obtained from the Juru and Jejawi Rivers in terms of the selected heavy metals' concentrations, whereas the alternative hypothesis states that at least one heavy metal concentration is different. The null and alternative hypotheses as defined in (5.9) are:

$$H_0: \boldsymbol{\mu_1} = \boldsymbol{\mu_2} \text{ vs } H_1: \boldsymbol{\mu_1} \neq \boldsymbol{\mu_2}$$

First, the variance-covariance matrix should be calculated for the two locations. The mean vector and variance-covariance matrices for the arsenic and heavy metal in cockles obtained from the Juru and Jejawi Rivers with Hotelling's test (T^2) are calculated using the built-in functions in R. The data are stored in the .CSV files `Example5_12a` and `Example5_12b` for the Juru and Jejawi Rivers, respectively. The R commands for testing the mean vectors of the arsenic and heavy metals in cockles obtained from the Juru and Jejawi Rivers are given below.

TABLE 5.6 The Arsenic and Heavy Metals Contents in Cockles Obtained From 20 Sites at the Juru and Jejawi Rivers (mg/L)

Location	Cr	As	Cd	Zn	Cu	Pb	Hg
Juru	0.20	2.38	0.82	0.23	0.20	0.13	1.40
Juru	0.17	2.94	0.86	0.23	0.16	0.13	1.45
Juru	0.15	2.53	0.88	0.21	0.24	0.11	1.31
Juru	0.15	2.61	0.92	0.29	0.20	0.13	1.19
Juru	0.15	2.89	0.92	0.23	0.21	0.14	1.46
Juru	0.17	2.58	0.89	0.21	0.20	0.12	1.41
Juru	0.18	2.83	0.88	0.21	0.20	0.12	1.30
Juru	0.18	2.61	0.89	0.20	0.18	0.11	1.28
Juru	0.16	2.64	0.90	0.24	0.18	0.12	1.27
Juru	0.18	2.66	0.89	0.21	0.20	0.12	1.33
Juru	0.18	2.73	0.90	0.19	0.18	0.13	1.33
Juru	0.16	2.57	0.89	0.18	0.17	0.12	1.42
Juru	0.16	2.61	0.84	0.22	0.17	0.12	1.28
Juru	0.17	2.86	0.86	0.25	0.17	0.12	1.27
Juru	0.16	2.60	0.88	0.20	0.18	0.11	1.24
Juru	0.17	2.83	0.91	0.19	0.19	0.12	1.42
Juru	0.16	2.83	0.91	0.23	0.17	0.12	1.31
Juru	0.17	2.62	0.88	0.23	0.16	0.13	1.36
Juru	0.14	2.42	0.90	0.20	0.17	0.11	1.36
Juru	0.18	2.69	0.92	0.19	0.17	0.12	1.24
Jejawi	0.17	2.61	0.85	0.20	0.17	0.12	1.36
Jejawi	0.16	2.90	0.88	0.18	0.21	0.13	1.44
Jejawi	0.15	2.57	0.89	0.22	0.22	0.12	1.41
Jejawi	0.16	2.70	0.84	0.19	0.19	0.14	1.28
Jejawi	0.15	2.77	0.88	0.17	0.20	0.12	1.35
Jejawi	0.19	2.61	0.84	0.19	0.15	0.12	1.41
Jejawi	0.18	2.69	0.88	0.20	0.18	0.11	1.44
Jejawi	0.17	2.75	0.88	0.18	0.23	0.12	1.41
Jejawi	0.17	2.65	0.88	0.22	0.19	0.12	1.36
Jejawi	0.17	2.60	0.82	0.22	0.19	0.12	1.44
Jejawi	0.16	2.70	0.93	0.20	0.20	0.12	1.25
Jejawi	0.16	2.43	0.88	0.20	0.19	0.12	1.44
Jejawi	0.16	2.69	0.93	0.18	0.20	0.12	1.40
Jejawi	0.18	2.82	0.89	0.19	0.18	0.12	1.23
Jejawi	0.15	2.60	0.84	0.20	0.20	0.14	1.29
Jejawi	0.18	2.88	0.86	0.19	0.17	0.12	1.21
Jejawi	0.15	2.90	0.89	0.18	0.17	0.12	1.36
Jejawi	0.17	2.89	0.87	0.20	0.17	0.13	1.32
Jejawi	0.18	2.45	0.79	0.24	0.15	0.14	1.50
Jejawi	0.15	2.61	0.85	0.18	0.19	0.11	1.32

```
#calculate the mean vector
m1 <- sapply (Example5_12a, mean)
m1
m2 <- sapply (Example5_12b, mean)
m2
round (cov(Example5_12a), digits = 3)
round (cov(Example5_12b), digits = 3)
# calculate Sp
n1 = length (Example5_12a $ Cu)
n2 = length (Example5_12b $ Cu)
sp = (((n1 - 1)*cov(Example5_12a)) + ((n2 - 1)*cov(Example5_12b)))/(n1 + n2 - 2)
round (sp, digits = 3)
dif = m1 - m2                  # calculate the difference
dif1 = t(dif)              # transpose
INV = solve(sp)            # calculate the inverse
HOT = ((n1*n2) / (n1 + n2))*(dif1%*%INV%*%dif)
HOT
```

The outputs from applying R commands to perform Hotelling's test for the arsenic and heavy metal concentrations in cockles are given below.

```
> #calculate the mean vector
> m1 <- sapply (Example5_12a, mean)
> m1
    Cr      As      Cd      Zn      Cu      Pb      Hg
0.1670  2.6715  0.8870  0.2170  0.1850  0.1215  1.3315
> m2 <- sapply (Example5_12b, mean)
> m2
    Cr      As      Cd      Zn      Cu      Pb      Hg
0.1655  2.6910  0.8685  0.1965  0.1875  0.1230  1.3610
> round (cov(Example5_12a), digits = 3)
    Cr     As      Cd      Zn      Cu Pb      Hg
Cr  0  0.000   0.000   0.000   0.000   0  0.000
As  0  0.023   0.001   0.000  -0.001   0  0.002
Cd  0  0.001   0.001   0.000   0.000   0  0.000
Zn  0  0.000   0.000   0.001   0.000   0 -0.001
Cu  0 -0.001   0.000   0.000   0.000   0  0.000
Pb  0  0.000   0.000   0.000   0.000   0  0.000
Hg  0  0.002   0.000  -0.001   0.000   0  0.006
> round (cov(Example5_12b), digits = 3)
    Cr     As      Cd      Zn Cu Pb      Hg
Cr  0  0.000   0.000   0.000   0   0  0.000
As  0  0.019   0.002  -0.001   0   0 -0.005
Cd  0  0.002   0.001   0.000   0   0 -0.001
Zn  0 -0.001   0.000   0.000   0   0  0.000
Cu  0  0.000   0.000   0.000   0   0  0.000
Pb  0  0.000   0.000   0.000   0   0  0.000
Hg  0 -0.005  -0.001   0.000   0   0  0.006
> # calculate Sp
> n1 = length (Example5_12a $ Cu)
```

(Continued)

(Continued)

```
> n2 = length (Example5_12b $ Cu)
> sp = (((n1 - 1)*cov(Example5_12a)) + ((n2 - 1)*cov(Example5_12b)))/(n1 + n2 - 2)
> round (sp, digits = 3)
     Cr    As     Cd Zn Cu Pb     Hg
Cr   0  0.000  0.000  0  0  0  0.000
As   0  0.021  0.001  0  0  0 -0.002
Cd   0  0.001  0.001  0  0  0  0.000
Zn   0  0.000  0.000  0  0  0  0.000
Cu   0  0.000  0.000  0  0  0  0.000
Pb   0  0.000  0.000  0  0  0  0.000
Hg   0 -0.002  0.000  0  0  0  0.006
> dif = m1 - m2                      # calculate the difference
> dif1 = t(dif)                      # transpose
> INV = solve(sp)                    # calculate the inverse
> HOT = ((n1*n2) / (n1 + n2))*(dif1%*%INV%*%dif)
> HOT
         [,1]
[1,] 21.63186
```

The first and second rows tell us that the mean vector for the arsenic and heavy metal concentrations is calculated using the function $\texttt{sapply()}$. Then we calculate the variance-covariance matrices. The Hotelling's test statistic is calculated after preparing all of the entries for the formula in (5.10). The null hypothesis is rejected because the calculated test statistic value (Hotelling's test) $T^2 = 21.63186$ is greater than the critical value $T^2_{0.05,7,38} = 19.26$. Thus it is clear that the two rivers are different in terms of their arsenic and heavy metal concentrations. Moreover, the average concentration of one or more parameters is different in cockles obtained from the two rivers.

Example 5.13: Physicochemical analysis of ripe banana pulp—A researcher is interested in comparing Cavendish and Dream bananas based on certain physicochemical properties. Three physicochemical properties were selected: the pH, TSS (° Brix), and WHC (Water Holding Capacity, g water/g dry sample) at 40°C, 60°C, and 80°C. Twelve samples were selected from each type of banana and tested for the selected parameters. Assume that the two populations are normally distributed. The results are given in Table 5.7. At $\alpha = 0.05$, is there sufficient evidence to conclude that the physicochemical properties of Dream and Cavendish bananas are different?

The null and alternative hypotheses for the mean vectors of Cavendish $\overline{Y}_1$ and Dream $\overline{Y}_2$ bananas, as defined in (5.9), are as follows:

$$H_0\text{: } \pmb{\mu_1} = \pmb{\mu_2} \text{ vs } H_1\text{: } \pmb{\mu_1} \neq \pmb{\mu_2}$$

$$H_0\text{:} \begin{bmatrix} 5.1267 \\ 4.2583 \\ 1.3742 \\ 1.7067 \\ 4.6660 \end{bmatrix} = \begin{bmatrix} 4.5067 \\ 3.4042 \\ 0.8350 \\ 1.0058 \\ 2.9617 \end{bmatrix} \text{ vs } H_1\text{:} \begin{bmatrix} 5.1267 \\ 4.2583 \\ 1.3742 \\ 1.7067 \\ 4.6660 \end{bmatrix} \neq \begin{bmatrix} 4.5067 \\ 3.4042 \\ 0.8350 \\ 1.0058 \\ 2.9617 \end{bmatrix}$$

The R commands used in Example 5.12 can be used to carry out Hotelling's test to compare the physicochemical parameters of Cavendish and Dream bananas. The data have been stored in the .CSV files $\texttt{Example5\_13a}$ and $\texttt{Example5\_13b}$ for Cavendish and Dream bananas, respectively.

TABLE 5.7 The Physicochemical Properties of Bananas

Variety	pH	TSS (°Brix)	WHC (g water/g dry sample)		
			40°C	60°C	80°C
Cavendish	5.60	4.33	1.53	1.61	4.86
Cavendish	5.57	4.03	1.50	1.74	4.04
Cavendish	4.76	3.77	1.52	1.81	4.98
Cavendish	5.56	4.10	1.69	1.76	5.03
Cavendish	4.95	3.97	1.42	1.77	4.04
Cavendish	4.84	4.40	1.28	1.56	5.01
Cavendish	5.07	4.50	1.25	1.68	5.01
Cavendish	4.94	4.43	1.08	1.68	4.84
Cavendish	5.04	4.30	1.54	1.88	4.83
Cavendish	4.93	4.57	1.18	1.58	3.99
Cavendish	5.05	4.30	1.26	1.78	4.77
Cavendish	5.21	4.40	1.24	1.63	4.59
Dream	4.31	3.67	0.82	1.05	2.81
Dream	4.41	3.80	0.80	1.04	2.87
Dream	4.35	3.00	0.81	0.98	2.96
Dream	4.49	3.40	0.77	1.04	2.97
Dream	4.39	3.67	0.68	0.88	2.70
Dream	4.43	3.33	0.80	0.99	3.05
Dream	4.44	3.47	0.72	0.87	3.21
Dream	4.44	3.57	0.88	1.02	2.90
Dream	4.52	3.20	1.07	1.00	3.10
Dream	4.79	3.17	0.97	1.06	3.00
Dream	4.68	3.27	0.81	1.06	2.98
Dream	4.83	3.30	0.89	1.08	2.99

```
#calculate the mean vector
m1 <- sapply (Example5_13a, mean)
m1
m2 <- sapply (Example5_13b, mean)
m2
round (cov(Example5_13a), digits = 3)
round (cov(Example5_13b), digits = 3)
# calculate Sp
n1 = length (Example5_13a $ pH)
n2 = length (Example5_13b $ pH)
sp = (((n1 - 1)*cov(Example5_13a)) + ((n2 - 1)*cov(Example5_13b)))/(n1 + n2 - 2)
round (sp, digits = 3)
```

(Continued)

(Continued)

```
dif = m1 - m2                    # calculate the difference
dif1 = t(dif)              # transpose
INV = solve (sp)           # calculate the inverse
HOT = ((n1*n2)/(n1 + n2))*(dif1%*%INV%*%dif)
HOT
```

The results (output) of performing Hotelling's test regarding the physicochemical parameters of bananas using R commands are given below.

```
> #calculate the mean vector
> m1 <- sapply (Example5_13a, mean)
> m1
      pH        TSS        X40C       X60C       X80C
5.126667   4.258333   1.374167   1.706667   4.665833
> m2 <- sapply (Example5_13b, mean)
> m2
      pH        TSS        X40C       X60C       X80C
4.506667   3.404167   0.835000   1.005833   2.961667
> round (cov(Example5_13a), digits = 3)
        pH     TSS    X40C    X60C    X80C
pH    0.087  -0.004  0.029  -0.001  -0.007
TSS  -0.004   0.058  -0.029 -0.015   0.007
X40C  0.029  -0.029  0.034   0.009   0.010
X60C -0.001  -0.015  0.009   0.010   0.003
X80C -0.007   0.007  0.010   0.003   0.165
> round (cov(Example5_13b), digits = 3)
        pH     TSS    X40C    X60C    X80C
pH    0.029  -0.018  0.008   0.006   0.007
TSS  -0.018   0.057  -0.012 -0.003  -0.017
X40C  0.008  -0.012  0.011   0.004   0.005
X60C  0.006  -0.003  0.004   0.005  -0.001
X80C  0.007  -0.017  0.005  -0.001   0.018
> # calculate Sp
> n1 = length (Example5_13a $ pH)
> n2 = length (Example5_13b $ pH)
> sp = (((n1 - 1)*cov(Example5_13a)) + ((n2 - 1)*cov(Example5_13b)))/(n1 + n2 - 2)
> round (sp, digits = 3)
        pH     TSS    X40C    X60C    X80C
pH    0.058  -0.011  0.018   0.002   0.000
TSS  -0.011   0.057  -0.021 -0.009  -0.005
X40C  0.018  -0.021  0.022   0.007   0.007d
X60C  0.002  -0.009  0.007   0.007   0.001
X80C  0.000  -0.005  0.007   0.001   0.092
> dif = m1 - m2      # calculate the difference
> dif1 = t(dif)      # transpose
> INV = solve (sp)   # calculate the inverse
> HOT = ((n1*n2)/(n1 + n2))*(dif1%*%INV%*%dif)
> HOT
    [,1]
[1,] 1032.232
```

The null hypothesis is rejected because the calculated test statistic value (Hotelling's test) `HOT1032.232` is greater than the critical value $T_{0.05,5,22}^2 = 16.945$. Thus Cavendish and Dream bananas have different physicochemical properties. It can be concluded that at least one of the selected physicochemical parameters is different for Cavendish and Dream bananas.

FURTHER READING

Abbas, F. M. A., Foroogh, B., Liong, M. T., & Azhar, M. E. (2008). Multivariate statistical analysis of antioxidants in dates (*Phoenix dactylifera*). *International Food Research Journal, 15*, 193−200.

Alkarkhi, A. F. M., Ismail, N., & Easa, A. M. (2008). Assessment of arsenic and heavy metal contents in cockles (*Anadara granosa*) using multivariate statistical techniques. *Journal of Hazardous Materials, 150*, 783−789.

Alkarkhi, A. F. M., & Low, H. C. (2012). *Elementary statistics for technologist*. Pulau Pinang, Malaysia: Universiti Sains Malaysia press.

Alkarkhi, A. F. M., Saifullah, B. R., Yeoh, S. Y., & Azhar, M. E. (2011). Comparing physicochemical properties of banana pulp and peel flours prepared from green and ripe fruits. *Food Chemistry, 129*, 312−318.

Allan, G. B. (2007). *Elemantary statistics: A step by step approach*. McGraw-Hill.

Bhabuk, U.R. (2013). *RG#47: Shaded Normal Curve* [Online]. Available: <http://rgraphgallery.blogspot.my/2013/04/shaded-normal-curve.html> Accessed 07.07.15.

Daniel, H. (2013). *Blog Archives, High Resolution Figures in R* [Online]. Available: <https://www.r-bloggers.com/author/daniel-hocking/> Accessed 15.07.17.

Datacamp. *Tutorial on the R Apply Family, The Apply Functions As Alternatives To Loops* [Online]. DataCamp. Available: <https://www.datacamp.com/community/tutorials/r-tutorial-apply-family#gs.rqdg = Dw> Accessed 08.09.17.

Donald, H. S., & Robert, K. S. (2000). *A first course*. McGraw-Hill.

Heuristic, A. (2011). *Paired sample t-test in R* [Online]. Available: <https://heuristically.wordpress.com/2011/09/28/paired-sample-t-test-in-r/> Accessed 19.09.16.

Mario, F. T. (2004). *Elementary statistics*. Wesky, Person Addison.

Robert, I.K. (2014). *Quick R, accessing the power of R* [Online]. Available: <http://www.statmethods.net/stats/ttest.html> Accessed 07.07.16.

Seow, E. K., Al-Karkhi, A. F. M., Mohd, K. M. T., Azhar, M. E., & Cheng, L. H. (2015). An investigation of potential fraud in commercial orange juice products in Malaysian market by cluster analysis and principal component analysis. *Malaysian Journal of Analytical Sciences, 19*, 377−387.

Slawa, R. (2013). *Normal distribution functions* [Online]. R-Bloggers. Available: <https://www.r-bloggers.com/normal-distribution-functions/> Accessed 13.09.17.

University of Cincinnati. *UC Business Analytics R Programming Guide* [Online]. Available: <http://uc-r.github.io/t_test> Accessed 15.06.17.

William, B.K. (2016). *Tutorials* [Online]. Available: <http://ww2.coastal.edu/kingw/statistics/R-tutorials/index.html> Accessed 05.08.16.

Chapter 6

Comparing Several Population Means

Learning outcomes

At the end of this chapter, you should be able

- *To describe the analysis of variance (ANOVA).*
- *To perform one-way ANOVA.*
- *To perform two-way ANOVA.*
- *To describe multivariate analysis of variance (MANOVA).*
- *To perform one-way MANOVA.*
- *To perform two-way MANOVA.*
- *To apply R commands for ANOVA and MANOVA.*
- *To understand and interpret the R output for ANOVA and MANOVA.*
- *To draw useful conclusions and report on them.*

6.1 INTRODUCTION

Hypothesis testing for a single variable, such as the t-test and Z-test, was studied in Chapter 5, Hypothesis Testing. These tests are used for testing one or two samples in a univariate case, while in a case with several variables (the multivariate case), Hotelling's test (T^2) is used. These tests cannot be used for three or more samples. Thus a new technique should be considered for this purpose. In this chapter, an analysis of variance (ANOVA) for one dependent variable and a multivariate analysis of variance (MANOVA) for several dependent variables measured on each sampling unit will be used to test the equality of means for three or more samples.

6.2 ANOVA AND MANOVA IN R

R supplies a variety of built-in functions to perform comparisons between population means. The functions cover the analysis of variance (ANOVA) (when there is only one dependent variable), the multivariate analysis of variance (MANOVA) (when there are several dependent variables), and the interaction plots between different variables. The following functions are used to carry out ANOVA and MANOVA in R, and we give the structure of each function.

1. The function `aov()` is used to perform one-way analysis of variance (ANOVA).

```
aov (response ~ independent variable, data = data frame)
aov (y ~ x, data = data frame)
```

where `y` represents the dependent variable and `x` represent the independent variable.

2. The function `aov.out()` is also used to carry out ANOVA. The functions `aov ()` and `aov.out()` are equivalent.

```
aov.out (Y ~ X, data = data frame)
```

Easy Statistics for Food Science with R. DOI: https://doi.org/10.1016/B978-0-12-814262-2.00006-6

3. The function `summary()`is used to provide more details on the analysis, namely, a p-value and a complete ANOVA Table.

```
summary (data frame)
```

4. Two-way ANOVA includes two cases: whereas there is no two-factor interaction between the different factors in the first case, we have this interaction in the second case. The R statistical software covers the two cases in the basic R installation. The `aov()` function is also used to perform two-way ANOVA. The R commands for two-way ANOVA with or without interactions are shown below.

```
aov (response ~ independent variables, data = data frame)
aov (y ~ x1 + x2, data = data frame)
```

Moreover, the commands for ANOVA, including the main effect and the effects of two-factor interactions, are given below.

```
aov (y ~ x1 + x2 + x1:x2, data = data frame)
or
aov (y ~ x1*x2, data = data frame)
```

In the above expressions, `y` represents the dependent variable (response) and `x1` and `x2` represent the independent variable.

5. The function `manova()` is used to conduct one-way MANOVA in R. Performing MANOVA in R requires the user to install and load the two packages `car` and `mvtnorm`. Three functions are needed to carry out MANOVA in R: `cbind ()`, `manova()`, and `summary()`. Let `x` represent the independent variable and let `y1, y2, ..., yk` be k dependent variables (responses).

```
y < - cbind (y1, y2, ..., yk), #y1,..., yk dependent vs.
MA < - manova (y ~ x) # x independent v.(factor)
```

The first function `cbind()` is a call to combine all of the responses and store them in a new place called `y`. The second function `manova(y ~ x)` is a call to perform MANOVA.

The third command is the `summary()` function. The `summary()` function is used with MANOVA to provide details on the analysis, including the degrees of freedom, the approximate F value, the p-value, and the test used to make a decision.

```
summary (MA, test = "wilks")
summary (MA, "wilks")
```

The test options include the "Hotelling-Lawley", "pillai", and "Roy" tests.

6. The function `manova()`is also used to execute two-way MANOVA, including the interactions. Let `x1` and `x2` be two independent variables and let `y1, y2, ..., yk` be k dependent variables (responses).

```
y <- cbind(y1, y2, ..., yk)
MA <- manova(y ~ x1*x2)
summary(MA, "wilks")
```

The test options include the "Hotelling-Lawley", "pillai", and "Roy" tests.

6.3 ANALYSIS OF VARIANCE (ANOVA)

The effect of one or more independent variables (factors) at more than two levels (settings, groups, samples) on a response (dependent variable) cannot be tested by using the Z or t tests as long as the independent variable has more than two levels. For example, a researcher wants to study the effects of temperature at three different settings (levels): 20°C, 30°C, and 40°C. Here, there are three groups (levels), and thus a new technique, analysis of variance (ANOVA), should be used to test the effect of three or more groups.

Analysis of variance (ANOVA) is an important technique used to analyze the variation of a response variable measured at different levels (groups) of one or more independent variables. The idea behind ANOVA is to partition the total variation into two sources of variation, one source attributed to the independent variable (variables), which represents the effect of an independent variable, and another source attributed to an unknown source, which is the error.

Note
- ANOVA can be used in the case of two means (two samples), and the results will be the same as the Z or t tests.

Assumptions of ANOVA
Three assumptions must be satisfied when conducting an analysis of variance. These assumptions are normality, independence, and homogeneity.

Normality: the samples must be selected from populations that are normally or approximately normally distributed.
Independence: the samples must be independent.
Homogeneity: the variances of the populations must be equal.

6.3.1 One-Way ANOVA

One-way ANOVA considers only one independent variable (X) (called a factor) at g levels (groups), and the objective is to study the effect of different levels of the factor on a continuous response (Y). For example, a researcher wants to study the effect of pressure on the yield extracted as a response; three different settings of pressure were chosen: 30, 40, and 50 MPa. Based on ANOVA, the variation or the changes in the yield extracted could be due to the effect of pressure or could be caused by an unknown source (error). Thus the total sum of squares is decomposed into two components, the sum of squares due to the independent variable pressure (SS_B), which is defined between groups (different levels), and the sum of squares due to error (SS_E). The total sum of squares can be written as:

$$SS_{Total} = SS_{Between\ groups} + SS_{Error}$$

$$SS_T = SS_B + SS_E$$

where SS_T represents the total sum of squares,

SS_E represents the sum of squares of error, and
SS_B represents the sum of squares between different levels (groups).

Each component is associated with a number of degrees of freedom.
The effect of a factor on a response variable is tested by the F test to decide whether or not the factor influences the response (statistically significant or not), as given in (6.1).

$$F = \frac{\text{Variance between groups}}{\text{Variance due to error}} = \frac{MS_B}{MS_E} \tag{6.1}$$

The general arrangement for a single-factor experiment is given in Table 6.1.
Notation and computational formulas

TABLE 6.1 The Arrangement of a Single-Factor Experiment

Factor (level)	Replication			
1	Y_{11}	Y_{12}	....	Y_{1n1}
2	Y_{21}	Y_{22}	...	Y_{2n2}
⋮	⋮	⋮	...	⋮
g	Y_{g1}	Y_{g2}	...	Y_{gng}

TABLE 6.2 The Entries of ANOVA Table

Source of variation (S.O.V)	Degrees of freedom (d.f)	Sum of squares (S.S)	Mean sum of squares (M.S)	F-value
Between groups	$g-1$	$SS_B = \sum n_i(\bar{y}_i - \bar{y})^2$	$MS_B = \dfrac{SS_B}{g-1}$	$F = \dfrac{MS_B}{MS_E}$
Within groups (Error)	$N-g$	$SS_{Error} = SS_E = \sum (n_i - 1)S_i^2$	$MS_E = \dfrac{SS_E}{N-g}$	
Total	$N-1$	$SS_T = \sum (y_i - \bar{y})^2$		

The notation used in the computational formula is given below:

g represents the number of groups (or levels),
n_i represents the number of observations in each group, $i = 1, 2, \ldots, g$,
$N = \sum_i n_i$ represents the total number of observations,
y_{ij} represents the observation j in the ith group, $j = 1, 2, \ldots, n_i$,
$\bar{y} = \dfrac{\sum y}{N}$ represents the grand mean, and $\bar{y}_i$ is the mean of group i.

The formulas for constructing a one-way ANOVA are summarized in Table 6.2.
Hypothesis testing for a one-way ANOVA
In a one-way ANOVA, researchers are interested in testing the following hypotheses:

$$H_0: \mu_1 = \mu_2 = \ldots = \mu_g$$

vs

$$H_1 : \mu_i \neq \mu_j \text{ for at least one pair}(i, j), \quad i = 1, 2, \ldots, g$$

For example, the null and alternative hypotheses for testing three different settings of pressure, 30, 40, and 50, on the yield extracted are

$$H_0: \mu_1 = \mu_2 = \mu_3$$

vs

$$H_1: \mu_i \neq \mu_j \text{ for at least one mean is different from others, } i = 1, 2, 3$$

The null hypothesis states that there is no significant difference in the yield extracted at 30, 40, and 50, while the alternative hypothesis states that at least one level (setting) of pressure out of three levels will give a different percentage of yield extracted.

Interpretation of the ANOVA Table

The interpretation of the ANOVA Table depends on the summary statistics provided by the last two columns in Table (6.2), namely, the F-value and p-value. The F-value in an ANOVA Table is called the calculated F-value (or computed value), and we should compare it with the theoretical value (critical value). If the calculated value is greater than the critical value at a specified significance level (alpha = 0.05, 0.01 or other values), then the null hypothesis is rejected. Computer outputs provide another value called the p-value (probability value), which is used to make a decision regarding the null hypothesis. If the p-value is small (for example, $p < 0.05$), then the null hypothesis is rejected.

Example 6.1: Estimated glycemic index (GI) for noodles—The estimated GI of yellow noodles prepared by partial substitution of wheat flour with banana pulp or peel flour was assessed. Nine types of noodles were investigated: conventional noodles, Ripe Cavendish pulp (RCPu), Ripe Cavendish peel (RCPe), Green Cavendish pulp (GCPu), Green Cavendish peel (GCPe), Ripe Dream pulp (RDPu), Ripe Dream peel (RDPe), Green Dream pulp (GDPu), and Green Dream peel (GDPe) noodles. The estimated glycemic index (GI is a ranking of carbohydrates on a scale from 0 to 100) of the noodles was calculated. Three replicates were used to run the experiment. Assume that the data are normally distributed. The data are given in Table 6.3.

Test the hypothesis that there is no difference among the means of GI values produced by various types of noodles (there is no effect of the type of noodle on the GI value). Use $\alpha = 0.05$.

There is only one independent variable (factor) of interest which is types of noodles, and the objective is to determine whether or not different types of noodles would result in different GI values. The null and alternative hypotheses are:

$$H_0: \mu_1 = \mu_2 = \ldots = \mu_9$$

vs

$$H_1 : \mu_i \neq \mu_j \text{ for at least one mean is different from others, } i = 1, 2, \ldots, 9$$

The null hypothesis states that all types of noodles produce the same value of GI and that the difference in the type of noodle does not affect the GI value. In contrast, the alternative hypothesis states that at least one type of noodle among the selected group of noodles would produce a different value of GI than the others. The equality of means for the GI values produced by various types of noodles can be tested by performing analysis of variance (ANOVA). The data for GI produced by different types of noodles were analyzed using the R statistical software. Two equivalent functions can be used to carry out ANOVA: `aov()` and `aov.out()`. The data had been stored as a .CSV (Example6_1) file.

```
aov (GI ~ Type.of.Noodles, data = Example6_1)
aov.out (GI ~ Type.of.Noodles, data = Example6_1)
```

TABLE 6.3 The Results for the Glycemic Index (GI) for Different Types of Noodles

Types of noodles	GI		
1	52.83	53.02	52.85
2	48.80	49.05	48.65
3	50.66	50.64	50.95
4	48.41	47.87	48.63
5	51.49	51.41	48.09
6	47.99	47.73	49.36
7	49.60	49.36	49.09
8	47.26	47.88	47.70
9	52.52	52.55	52.54

The two commands provide information on the sum of squares and degrees of freedom without detailed information on the analysis such as the p-value and F value. Thus the function summary() should be used to obtain detailed information with a complete ANOVA Table, including the degrees of freedom (Df), sum of squares (Sum Sq), mean sum of squares (Mean Sq), F value, and p-value (Pr(> F).

```
summary (aov)
or
summary (aov.out)
```

The function aov() is used to conduct analysis of variance (ANOVA) for the GI produced by different types of noodles.

```
av < - aov(GI ~ Type.of.Noodles, data = Example6_1)
av
summary (av)
```

The results of using the aov() and summary() functions to perform ANOVA are given below.

```
> av < - aov(GI ~ Type.of.Noodles, data = Example6_1)
> av
Call:
   aov(formula = GI ~ Type.of.Noodles, data = Example6_1)
Terms:
                 Type.of.Noodles  Residuals
Sum of Squares           85.34336    9.86707
Deg. of Freedom                 8         18

Residual standard error: 0.7403853
Estimated effects may be unbalanced
> summary (av)
                 Df Sum Sq Mean Sq F  value      Pr( >F)
Type.of.Noodles   8  85.34  10.668  19.46   2.26e-07 ***
Residuals        18   9.87   0.548
---
Signif. codes: 0 '***' 0.001 '**' 0.01 '*' 0.05 '.' 0.1 ' ' 1
```

The results of R commands include a call to aov() to perform ANOVA on the data (which had been stored as the . CSV file Example6_1) and save the result in a new place called av. Then there is a call to print the results. The results of using the aov() function are the sum of squares, degrees of freedom (Deg. of Freedom), residuals, and residual standard error. The summary() function provides more details on the entries of the ANOVA Table, including the degrees of freedom (Df), the sum of squares (sum sq), the mean sum of squares (Mean sq), the F value, and the p-value (pr (>F)).

Based on the ANOVA results, it is clear that the type of noodle has a significant effect on the GI value as indicated by the Pr(>F) column, which represents the p-value of 2.26e-07. Because this value is very small, the null hypothesis of equal means of GI produced by different types of noodles is rejected. As a result, at least one type of noodle would produce a different GI value compared to the other types. Thus the selection of a type of flour must account for the requirement of having a proper GI value when producing noodles. This result is general, and to this point, we cannot identify which type of flour results in a different GI. Further analysis should be undertaken to determine the source of the differences. In addition, multiple comparisons should be used to determine the type of flour that results in a different GI value if required.

TABLE 6.4 The Data for the Total Flavonoid (TFC) of Dates (mg CEQ/ 100 g dw)

Type of date	TFC	Type of date	TFC
Honey	1.64	Piarom	3.03
Honey	1.73	Piarom	3.29
Honey	1.82	Piarom	3.90
Sahroon	0.96	Kabkab	1.01
Sahroon	0.98	Kabkab	1.01
Sahroon	1.63	Kabkab	1.25
Bam	1.64	Zahedi	3.58
Bam	1.77	Zahedi	4.71
Bam	1.96	Zahedi	3.36
Jiroft	0.85	Kharak	55.98
Jiroft	0.89	Kharak	72.31
Jiroft	1.04	Kharak	79.58

Example 6.2:Total flavonoid of dates—An experiment was conducted to analyze the edible parts of date palm (*Phoenix dactylifera*) fruits (DPF) for their antioxidant activities. The total flavonoid content (TFC) mg catechin equivalents (CEQ)/100 g of the DPFwas measured. Eight types of dates were chosen to represent soft, semidry, and dry dates. The samples include four types of soft dates (SD) (Bam dates, Honey dates, Jiroft dates, and Kabkab dates); three types of semidry dates (SDD) (Sahroon dates, Piarom dates, and Zahedi dates); and one type of dry date (DD) (Kharak dates). The data are given in Table 1.5 with the other antioxidant parameters. We assume that the data are normally distributed. Then we reproduce the data regarding the total flavonoid content (TFC) in Table 6.4.

At $\alpha = 0.05$, we test whether the total flavonoid content (TFC) in various types of datesis different or not. The data had been stored as a `.CSV (Example6_2)` file.

The independent variable (factor) that affects the value of the TFC in various types of dates is the type of dates used. The null and alternative hypotheses for the average values of the TFC produced by various types of dates are as follows:

$$H_0: \mu_1 = \mu_2 = \ldots = \mu_8$$

vs

$$H_1 : \mu_i \neq \mu_j \text{ for at least one mean is different from others}, i = 1, 2, \ldots, 8$$

The null hypothesis states that all types of dates result in the same total flavonoid content (TFC), which means that the type of date does not affect the TFC value. The alternative hypothesis states that at least one type of date would result in a different value of TFC than the others. The data in Table 6.4 for the total flavonoid content were analyzed using R's built-in functions to produce an ANOVA Table with detailed information to make a decision regarding the hypothesis.

The results of applying the `aov()` and `summary()` functions to carry out ANOVA to test the TFC in various types of dates are presented below.

```
> av <- aov(TFC ~ Type.of.date, data = Example6_2)
> av
Call:
     aov(formula = TFC ~ Type.of.date, data = Example6_2)
Terms:
                    Type.of.date   Residuals
Sum of Squares          11909.849    294.026
Deg. of Freedom             7            16

Residual standard error: 4.286795
Estimated effects may be unbalanced
> summary(av)
               Df   Sum Sq  Mean  Sq  F value    Pr(>F)
Type.of.date    7   11910   1701.4    92.58    9.55e-12 ***
Residuals      16     294     18.4
---
Signif. codes: 0 '***' 0.001 '**' 0.01 '*' 0.05 '.' 0.1 ' ' 1
```

The same results will be obtained if the function `aov.out()` is used.

```
av <- aov.out (TFC ~ Type.of.date, data = Example6_2)
av
summary(av)
```

The results of ANOVA showed that there is a significant difference in the TFC values in various types of dates, as the p-value is very small at `9.55e-12`. This finding leads us to reject the null hypothesis; hence the type of date affects the total flavonoid content (TFC) value.

In summary, it can be said that edible types of dates contain the different values of TFC. This distinction could be due to the type of date (e.g., SD, SDD or DD) that influences the TFC value, or it could be due to other reasons.

6.3.2 Two-Way ANOVA

The idea of one-way analysis of variance can be extended to include two independent variables (factors); each variable has at least two levels and one response variable. The technique for analyzing the effect of two independent variables is called a two-way analysis of variance.

For example, a researcher wants to investigate the effects of two factors on the weight of banana flour (g) reduction as a response: the temperature at 50, 60, and 70 and the type of acid (oxalic, citric, and combined). The variation in the weight of banana flour could be due to the temperature, the effect of the type of acid, or the interaction between the temperature and the type of acid, which are well-known sources of variation. The remainder of the variation could be due to the error (from an unknown source). Thus the total sum of squares (SS_T) is decomposed into four components: the sum of squares due to error (SS_E); the sum of squares due to the first factor, temperature (SS_A); the sum of squares due to the second factor, type of acid (SS_B); and the component due to the interaction between pH and cation concentration (SS_{AB}). An interaction exists between two factors if the effect of one of the factors changes for different levels of the other factors.

The total sum of squares for a two-way ANOVA, including the four components, can be written as

$$SS_T = SS_{Error} + SS_A + SS_B + SS_{AB}$$

where

SS_T represents the total sum of squares,
SS_E represents the sum of squares of error,

TABLE 6.5 General Arrangement for a Two-Factor Experiment

		Factor B			
		1	**2**	...	**q**
Factor A	1	$Y_{111}, Y_{112}, \ldots, Y_{11n}$	$Y_{121}, Y_{122}, \ldots, Y_{12n}$		$Y_{1q1}, Y_{1q2}, \ldots, Y_{1qn}$
	2	$Y_{211}, Y_{212}, \ldots, Y_{21n}$	$Y_{221}, Y_{222}, \ldots, Y_{22n}$	...	$Y_{2q1}, Y_{2q2}, \ldots, Y_{2qn}$
	⋮	⋮	⋮	⋮	⋮
	g	$Y_{q11}, Y_{q12}, \ldots, Y_{g1n}$	$Y_{g21}, Y_{g22}, \ldots, Y_{g2n}$	...	$Y_{gq1}, Y_{gq2}, \ldots, Y_{gqn}$

TABLE 6.6 The Entries Two-Way ANOVA

Source of variation S.O.V	Degrees of freedom (d.f)	Sum of squares (S.S)	Mean sum of squares (M.S)	F-value
A	$g-1$	$SS_A = \frac{1}{qn} \sum_{i=1}^{g} y_{i..}^2 - \frac{y_{...}^2}{pqn}$	$MS_A = \frac{SS_A}{g-1}$	$F = \frac{MS_A}{MS_E}$
B	$q-1$	$SS_B = \frac{1}{gn} \sum_{j=1}^{q} y_{.j.}^2 - \frac{y_{...}^2}{gqn}$	$MS_B = \frac{SS_B}{q-1}$	$F = \frac{MS_B}{MS_E}$
AB	$(g-1)(q-1)$	$SS_{AB} = SS_{subtotal} - SS_A - SS_B$ $$SS_{subtotal} = \frac{1}{n} \sum_{i=1}^{g} \sum_{j=1}^{q} y_{ij.}^2 - \frac{y_{...}^2}{pqn}$$	$MS_{AB} = \frac{SS_{AB}}{(g-1)(q-1)}$	$F = \frac{MS_{AB}}{MS_E}$
Error	$gq(n-1)$	$SS_{Error} = SS_E = SS_T - SS_{AB} - SS_A - SS_B$	$MS_E = \frac{SS_E}{gq(n-1)}$	
Total	$gqn-1$	$SS_T = \sum \sum \sum (\bar{y}_{i..} - \bar{y}_{...})^2 = \sum_{i=1}^{g} \sum_{j=1}^{q} \sum_{k=1}^{n} y_{ijk}^2 - \frac{y_{...}^2}{gqn}$		

SS_A represents the sum of squares between different levels of the first factor,
SS_B represents the sum of squares between different levels of the second factor, and
SS_{AB} represents the sum of squares due to the interaction between the two factors.

Each component is associated with a number of degrees of freedom.

The general arrangement for a two-factor experiment with factor A at g levels and factor B at q levels, is given in Table 6.5.

Notation and computational formulas

Sum of all the observations is $y_{...} = \sum_{i=1}^{g} \sum_{i=1}^{q} \sum_{k=1}^{n} y_{ijk}$

Sum of the q observations in row i is $y_{i..} = \sum_{j=1}^{q} \sum_{k=1}^{n} y_{ijk}$

Sum of the g observations in column j is $y_{.j.} = \sum_{i=1}^{g} \sum_{k=1}^{n} y_{ijk}$

Sum of each cell under the same combination is $y_{ij.} = \sum_{k=1}^{n} y_{ijk}$

The formulas for the two-way ANOVA entries are summarized in Table 6.6.

6.3.2.1   Hypothesis Testing for a Two-Way ANOVA

Suppose there are two factors, factor A at g levels and factor B at q levels. The researchers are interested in testing the following hypotheses:

1. Hypothesis about the effect of the first factor (factor A)

$$H_0: \alpha_1 = \alpha_2 = \ldots = \alpha_g = 0$$

The null hypothesis states that there is no effect of factor A on the response (the means of different levels are equal), while the alternative hypothesis is

$$H_1: \text{at least one } \alpha_i \neq 0 \quad i = 1, 2, \ldots, g$$

The alternative hypothesis states that there is an effect of factor A on the response.

2. Hypothesis about the effect of the second factor (factor B)

$$H_0: \beta_1 = \beta_2 = \ldots = \beta_q = 0$$

The null hypothesis states that there is no effect of factor B on the response. The alternative hypothesis is

$$H_1: \text{at least one } \quad \beta_j \neq 0 \, j = 1, 2, \ldots, q$$

The alternative hypothesis states that there is an effect of factor B on the response.

3. Hypothesis about the effect of interaction between the two factors A and B.

$$H_0: (\alpha\beta)_{ij} = 0 \text{ for all } i, j$$

The null hypothesis states that there is no interaction between factor A and factor B (both factors are independent). The alternative hypothesis is

$$H_1: \text{at least one } (\alpha\beta)_{ij} \neq 0$$

The alternative hypothesis states that there is an interaction between factor A and factor B.

Note

- It is recommended that the main effect should not be interpreted if the interaction exists. However, the main effect can be interpreted with some caution.

Example 6.3:Peroxide value of deep-fried snacks—An experiment was conducted to study the effect of type of snack and type of packaging on the peroxide value meq O2/kg (milliequivalents of peroxide per kg). Two different types of deep fried snacks (tapioca chips and fish crackers) and three types of packaging were chosen; the types of packaging are laminated plastic gas packaging (L), polypropylene (transparent) (p), and polyethylene (transparent) (t). Three replicates at each combination of the type of snack and the type of packaging were run. The data are given in Table 6.7.

TABLE 6.7 The Results of the Peroxide Value (meq/kg) for the Type of Snacks

Type	Packaging	Peroxide	Type	Packaging	Peroxide
Tapioca	L	4.63	Fish	L	0.61
Tapioca	L	3.45	Fish	L	0.60
Tapioca	L	4.22	Fish	L	0.60
Tapioca	t	2.70	Fish	t	1.50
Tapioca	t	2.66	Fish	t	1.48
Tapioca	t	2.59	Fish	t	1.51
Tapioca	P	17.05	Fish	p	1.39
Tapioca	P	17.12	Fish	p	1.31
Tapioca	P	16.97	Fish	p	1.28

Assuming that the data are normally distributed, conduct an analysis of variance and test it at $\alpha = 0.05$. Verify the hypothesis that there is no difference between the two types of snacks, no difference between three types of packaging, and no interaction between different types of snacks and types of packaging on the peroxide value.

This example studies the effects of two independent variables (factors), namely, the type of snack and the type of packaging, on the peroxide value (response). In addition to the main effects of the two factors, the interaction between the type of snacks and the type of packaging should be tested to ascertain whether the interaction exists (and whether it influences the peroxide value).

The results of the peroxide value obtained from tapioca chips and fish crackers were analyzed using two-way ANOVA to determine the effects of the selected factors. The built-in functions in R were used to calculate the entries of the two-way ANOVA.

```
Snack = as.factor (Example6_3 $ Snack)
Packaging = as.factor (Example6_3 $ packaging)
Peroxide = Example6_3 $ Peroxide
aov (Peroxide ~ Snack*Packaging)   # ANOVA with interaction
summary (aov(Peroxide ~ Snack*Packaging))
```

The first two lines were used to define and extract the two variables (the type of snacks and type of packaging) as a factor using the function `factor()` or `as.factor()`. The third line is used to define and extract the dependent variable (peroxide). The last two calls execute the ANOVA using the functions `aov()` and `summary()` to provide detailed information.

```
> Snack = as.factor (Example6_3 $ Snack)
> Packaging = as.factor (Example6_3 $ packaging)
> Peroxide = Example6_3 $ Peroxide
> aov (Peroxide ~ Snack*Packaging)       # ANOVA with interaction
Call:
    aov(formula = Peroxide ~ Snack * Packaging)
Terms:
                    Snack    Packaging   Snack:Packaging   Residuals
Sum of Squares   207.46845   194.78841      183.54443       0.74227
Deg. of Freedom        1           2              2            12

Residual standard error: 0.2487078
Estimated effects may be unbalanced
> summary (aov(Peroxide ~ Snack*Packaging))
                 Df  Sum Sq  Mean Sq  F value     Pr(>F)
Snack             1  207.47   207.47     3354   4.64e-16 ***
Packaging         2  194.79    97.39     1575   2.99e-15 ***
Snack:Packaging   2  183.54    91.77     1484   4.27e-15 ***
Residuals        12    0.74     0.06
  - - -
  Signif. codes: 0 '***' 0.001 '**' 0.01 '*' 0.05 '.' 0.1 ' ' 1
```

Here we examine the interactions between the types of snacks and the types of packaging. `Snack:Packaging` should be interpreted first. The results of the ANOVA showed a significant interaction between the types of snacks and the types of packaging, as the p-value is very small (p-value <4.27e-15 in the ANOVA Table under Pr(F)). This discovery means that the two factors do not work independently. A significant interaction indicates that the combination of the types of snacks and the types of packaging affects the value of the peroxide, and both factors should be studied regardless of the main effect (which may be significant or not significant). The main effect of the types of snacks was significant (the null hypothesis is rejected because the p-value <4.64e-16), and the main effect of the types of packaging was also significant (the null hypothesis is rejected because the p-value <2.99e-15).

The behavior of each variable in the presence of another variable can be studied by producing a graph of the interaction between types of snacks and types of packaging and observing the value of peroxide using R commands. The function `interaction.plot()` is used to generate the two-factor interaction between the types of snacks and the types of packaging.

```
Interaction.plot (x1, x2, Y)
```

The interaction.plot() functions require one to install the graphics package before running the command and producing the interaction graph in R.

Here X1 and X2 represent the independent variables and Y represents the dependent variable (response) (Fig. 6.1).

```
interaction.plot (Snack, Packaging, Peroxide)
```

Example 6.4: Weight of banana flour—An experiment was conducted to study the effect of the temperature and the type of acid on the weight of banana flour (g). Three settings of the temperature (50, 60, and 70) and three different types of acid were chosen (oxalic, citric, and combined). Two replicates at each combination of the temperature and the type of acid were used. The data obtained from the experiment are given in Table 6.8.

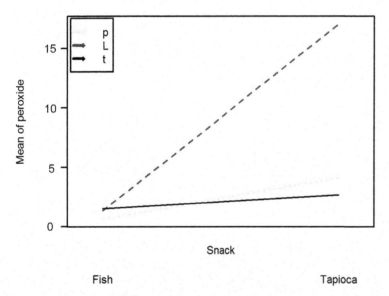

FIGURE 6.1 Interaction plot between types of snacks and types of packaging.

TABLE 6.8 The Weight of Banana Flour (g)

Temperature	Type of acid		
	Oxalic	Citric	Combined
50	93.320 92.956	93.332 93.053	93.320 92.956
60	84.521 91.143	91.686 90.584	90.814 91.050
70	92.837 92.515	92.301 92.601	92.396 92.494

Assuming that the weight is normally distributed with a common variance, we conduct an analysis of variance and test whether the temperature and type of acid influenced the weight of the banana flour at $\alpha = 0.05$. The data had been stored as a .CSV (Example6_4) file.

Two factors are of interest: the temperature and the type of acid. Each factor has three possible levels. The data should be analyzed with a two-way analysis of variance to test whether the temperature, the type of acid, and the interaction between the temperature and the type of acid affected the weight of banana flour. The following commands will produce the required entries for constructing a two-way ANOVA Table for the weight of banana flour in R.

```
Temperature = factor (Example6_4 $ Temperature, levels = c(50, 60, 70))
Acid = factor (Example6_4 $ Acid, levels = c("Oxalic", "Citric", "Combine"))
Weight = Example6_4 $ Weight
aov (Weight ~ Temperature*Acid)        # ANOVA with interaction
summary (aov(Weight ~ Temperature*Acid))
```

The first two commands (with factor) are used to define and extract the two variables of the temperature and acid as two factors (independent variables); the third command is to define the weight of banana flour as a response variable (dependent). The aov() function is used to conduct an analysis of variance, and summary() is used to give detailed information about the analysis.

The results of using R commands to build an ANOVA Table for the weight of banana flour are given below.

```
> Temperature = factor (Example6_4 $ Temperature, levels = c(50, 60, 70))
> Acid = factor (Example6_4 $ Acid, levels = c("Oxalic", "Citric", "Combine"))
> Weight = Example6_4 $ Weight
> aov (Weight ~ Temperature*Acid)        # ANOVA with interaction
Call:
   aov(formula = Weight ~ Temperature * Acid)
Terms:
                   Temperature        Acid Temperature:Acid Residuals
Sum of Squares     34.23266  4.02514         9.75551      22.83355
Deg. of Freedom           2         2               4             9

Residual standard error: 1.592816
Estimated effects may be unbalanced
> summary (aov(Weight ~ Temperature*Acid))
                Df  Sum Sq  Mean Sq  F value Pr(>F)
Temperature      2   34.23   17.116    6.747  0.0162 *
Acid             2    4.03    2.013    0.793  0.4816
Temperature:Acid 4    9.76    2.439    0.961  0.4735
Residuals        9   22.83    2.537
---
Signif. codes:  0 '***'  0.001  '**'  0.01  '*'  0.05  '.'  0.1 ' ' 1
```

To interpret the results properly, the interaction between the temperature and the type of acid should be tested first. The above results of the ANOVA showed that the interaction between the temperature and the type of acid was not significant, as the p-value is greater than 0.05 (Pr(>F), p-value <0.4732). Thus the null hypothesis regarding the interaction between the temperature and the type of acid is not rejected, which means that there is no interaction between the temperature and the type of acid, i.e., the two factors work independently. The second step is to test for the main effects of the temperature and the type of acid. The main effect of the temperature was significant because the p-value is small (p-value <0.0162) and less than 0.05, while the main effect of the type of acid was not significant (p-value <0.4816).

The next step is to interpret the main effects. The main effect of the temperature exhibited a significant effect on the weight of banana flour, which means that the temperature influences the weight of banana flour. In other words, a

TABLE 6.9 The Entries for a Three-Way ANOVA

Source of variation (S.O.V)	Degrees of freedom (d.f)	Sum of squares (S.S)	Mean sum of squares (M.S)	F-value
A	$a-1$	SS_A	MS_A	$F_A = SS_A/SS_E$
B	$b-1$	SS_B	MS_B	$F_B = SS_B/SS_E$
C	$c-1$	SS_C	MS_C	$F_C = SS_C/SS_E$
AB	$(a-1)(b-1)$	SS_{AB}	MS_{AB}	$F_{AB} = SS_{AB}/SS_E$
AC	$(a-1)(c-1)$	SS_{AC}	MS_{AC}	$F_{AC} = SS_{AC}/SS_E$
BC	$(b-1)(c-1)$	SS_{BC}	MS_{BC}	$F_{BC} = SS_{BC}/SS_E$
ABC	$(a-1)(b-1)(c-1)$	SS_{ABC}	MS_{ABC}	$F_{ABC} = SS_{ABC}/SS_E$
Error	$abc(n-1)$	SS_E	MS_E	
Total	$abcn-1$	SS_T		

different temperature setting would yield a different weight, and as a result, the researcher should choose the most suitable temperature to achieve the target of the research. However, the type of acid did not show a significant effect on the weight, which means that any type of acid would yield the same weight.

In summary, only the effect of the temperature should be considered in future work and any type of acid can be used.

Note
• An ANOVA Table is constructed in the same manner for k independent variables. For example, an ANOVA Table for three independent variables is given in Table 6.9.

6.4   MULTIVARIATE ANALYSIS OF VARIANCE (MANOVA)

The idea of a multivariate analysis of variance (MANOVA) is the same as a univariate analysis of variance (ANOVA), because both methods are used to test the equality of means for three or more samples. However, the difference between ANOVA and MANOVA is that ANOVA is used when only one response variable (dependent variable) is measured for each experimental unit, while MANOVA is used when several response variables (dependent variables) are measured for each experimental unit.

Assumptions for MANOVA

Normality: the samples must be selected from populations that are normally or approximately normally distributed.
Independence: the samples must be independent.
Homogeneity: all populations have a common variance-covariance matrix.
Multivariate normal: each population is multivariate normally distributed.

6.4.1   One-Way MANOVA

The concept of one-way MANOVA is the same as one-way ANOVA, and the difference between them is the number of response variables (dependent variables). In a multivariate case, there are g populations, and the objective is to compare these populations based on several measurements (dependent variables), i.e., whether the g samples' mean vectors are the same or not. A sample of size n is selected from k-variate normal populations with equal covariance matrices, as displayed in Table 6.10.

Hypothesis testing
In a one-way MANOVA, the researchers are interested in the following hypothesis:

$$H_0: \mu_1 = \mu_2 = \cdots = \mu_g$$

TABLE 6.10 The Arrangement of Multivariate Data for a One-Way MANOVA

Sample 1	Y_{11}	Y_{12}	$\cdots$	Y_{1n}
Sample 2	Y_{21}	Y_{22}	$\cdots$	Y_{2n}
$\vdots$	$\vdots$	$\vdots$	$\cdots$	$\vdots$
Sample g	Y_{g1}	Y_{g2}	$\cdots$	Y_{gn}

TABLE 6.11 The Entries of a One-Way MANOVA Table

Source of variation (S.O.V)	Matrix of sum of squares and cross products (SSP)	Degrees of freedom (d.f)
Between	$B = n \sum_{i=1}^{g} (\bar{y}_i - \bar{y})(\bar{y}_i - \bar{y})'$	$g - 1$
Error (residual)	$W = \sum_{i=1}^{g} \sum_{j=1}^{n} (y_{ij} - \bar{y}_i)(y_{ij} - \bar{y}_i)'$	$N - g$
Total	$B + W = \sum_{i=1}^{g} \sum_{j=1}^{n} (y_{ij} - \bar{y})(y_{ij} - \bar{y})'$	$N - 1$

$$H_0: \begin{bmatrix} \mu_{11} \\ \mu_{12} \\ \vdots \\ \mu_{1k} \end{bmatrix} = \begin{bmatrix} \mu_{21} \\ \mu_{22} \\ \vdots \\ \mu_{2k} \end{bmatrix} = \ldots = \begin{bmatrix} \mu_{g1} \\ \mu_{g2} \\ \vdots \\ \mu_{gk} \end{bmatrix}$$

The null hypothesis states that the g means are equal for each dependent variable.

$$H_1 : \text{at least two } \mu's \text{ are different}$$

where μ is a vector.

If the null hypothesis is not rejected, the g means are equal for each dependent variable, while rejecting the null hypothesis means that at least two means (groups) differ for at least one dependent variable.

Using a similar idea as that employed in univariate analysis to find the between sum of squares (B) and the within sum of squares (W), the two matrices with associated degrees of freedom are given in Table 6.11.

The null hypothesis is rejected if the likelihood ratio given in (6.2) is less than or equal to the critical value.

$$\Lambda = \frac{|W|}{|W + B|}, 0 \leq \Lambda \leq 1 \tag{6.2}$$

Reject if $\Lambda \leq \Lambda_{\alpha,k,g-1,N-g}$ (critical value)

This test is called the Wilk's test.

Note
- The hypothesis is rejected for small values of Λ.
- Other tests can be used to verify the null hypothesis for the multivariate case: Pillai, Lawley-Hotelling, and Roy's largest root. R provides commands to perform all multivariate test (MANOVA).

Example 6.5: Texture of banana flour—The effect of time on the texture of banana flour was investigated. Two responses (dependent variables) are of interest: the tensile strength (TS, kPa) and elasticity (kPa). Six time settings were selected: 5, 10, 15, 20, 25, and 30 minutes. In addition, three replicates were used to run the experiment. A researcher wants to know whether the time (min) influences the texture of banana flour. Assume that the data are normally distributed. The data are given in Table 6.12.

TABLE 6.12 The Results for the TS and Elasticity of Banana Flour

Time	T.S	Elasticity	Time	T.S	Elasticity
5	83.90	41.89	20	45.14	29.90
5	79.26	39.23	20	50.21	32.29
5	81.26	40.78	20	51.17	32.48
10	72.74	39.13	25	44.22	30.35
10	67.45	37.69	25	43.44	28.47
10	70.85	37.23	25	42.71	29.07
15	56.60	34.04	30	41.38	29.44
15	58.86	34.97	30	40.20	28.36
15	56.02	34.38	30	41.55	29.00

The texture of banana flour was assessed by studying the effect of time (the independent variable or factor) on both the tensile strength and the elasticity. The function `manova()` was used to conduct a one-way multivariate analysis of variance (MANOVA) to decide whether the time (independent variable) affects significantly the behavior of the tensile strength, the elasticity, or both. Furthermore, the function `cbind()` was used to prepare the data for the function `manova()`, and the last function is `summary()`, which provided more details about the analysis. The data for the texture of banana flour had been stored as a `.CSV(Example5_5)` file and analyzed using MANOVA.

```
TS = Example6_5 $ TS
Elasticity = Example6_5 $ Elasticity
Time = as.factor (Example6_5 $ Time)
Y <- cbind (TS, Elasticity)
Ma <- manova (Y ~ Time)
summary (Ma, test = "Wilks")
summary (Ma, test = "Hotelling-Lawley")
summary (Ma, test = "Roy")
summary (Ma, test = "Pillai")
```

The first three rows were used to define and extract the two responses (the tensile strength and elasticity) and the independent variable (time). The function `cbind()` was used to combine the two responses of the tensile strength (TS) and elasticity in one file and save it in a new place called Y. The data for the texture of banana flour were analyzed, and the function `manova()`was used to carry out MANOVA for the texture of banana flour and place the results in a new location called Ma. In addition, the function `summary()` was used after performing MANOVA to provide more details regarding the four multivariate tests (Wilks, Hotelling-Lawley, Pillai, Roy), the test statistic's value, and the p-value for each test.

```
> TS = Example6_5 $ TS
> Elasticity = Example6_5 $ Elasticity
> Time = as.factor (Example6_5 $ Time)
> Y <- cbind (TS, Elasticity)
> Ma <- manova (Y ~ Time)
> summary (Ma, test = "Wilks")
```

(Continued)

(Continued)

```
             Df    Wilks   approx F  num Df  den Df  Pr(>F)
Time         5 0.0072225   23.687      10      22    1.649e-09 ***
Residuals   12
---
Signif. codes: 0 '***' 0.001 '**' 0.01 '*' 0.05 '.' 0.1 ' ' 1
> summary(Ma, test = "Hotelling-Lawley")
           Df Hotelling-Lawley approx F num Df den Df  Pr(>F)
Time        5          85.371   85.371     10     20 3.635e-14 ***
Residuals  12
---
Signif. codes: 0 '***' 0.001 '**' 0.01 '*' 0.05 '.' 0.1 ' ' 1
> summary(Ma, test = "Roy")
           Df    Roy approx F num Df den Df     Pr(>F)
Time        5 84.756  203.41      5     12   3.631e-11 ***
Residuals  12
---
Signif. codes: 0 '***' 0.001 '**' 0.01 '*' 0.05 '.' 0.1 ' ' 1
> summary(Ma, test = "Pillai")
           Df Pillai approx F  num Df  den Df    Pr(>F)
Time        5  1.369   5.2066     10      24   0.0004498 ***
Residuals  12
---
Signif. codes: 0 '***' 0.001 '**' 0.01 '*' 0.05 '.' 0.1 ' ' 1
```

The MANOVA results from testing the effect of time on the tensile strength and elasticity showed that the null hypothesis, which states that the means are equal and there is no effect of time on the tensile strength or elasticity, is rejected since the p-value is very small. The p-value is presented in the output under pr(>F) for all of the multivariate tests (Wilks, p-value < 1.649e-09, Hotelling-Lawley, p-value < 3.635e-14, Roy, p-value < 3.631e-11, Pillai, p-value < 0.0004498). This result indicates that time has a significant effect on the strength and elasticity. In conclusion, time is an important factor that affects the texture of banana flour and should be considered an influential factor for any future activity.

Example 6.6: Quality of fish crackers—A researcher wants to investigate the quality of fish crackers obtained from different brands. To that end, five parameters were studied: the fat content, moisture content, and color content in terms of the redness (a*), yellowness (b*), and lightness (L*). The samples were selected from seven different brands, and three samples were selected from each brand. Assume that the data are normally distributed. The researcher was interested in testing the difference between the various brands regarding the fat, moisture, and color contents of the product. Use $\alpha = 0.05$ to conduct the analysis. The data are given in Table 6.13. Use MANOVA to test whether the brand of fish cracker influences the quality of the crackers.

The researcher is interested in measuring five parameters (responses) to decide whether the quality of fish crackers is significantly affected by the brands (the independent variables or factors), which would imply that each brand has a different quality of fish cracker. Alternatively, it is possible that the brand does not affect the quality of the fish crackers significantly, which would imply that all of these crackers have the same quality regardless of the brand. MANOVA was used to determine the effect of the brands on the quality of fish crackers using similar steps and commands to those in Example 6.5. The test statistic values for Wilk's, Hotelling's, and the Pillai and Roy tests using R were calculated for the parameters measured from samples of fish crackers. The data had been stored as a .CSV (Example6_6) file.

TABLE 6.13 The Selected Parameters for Fish Crackers

Market	Fat content (%)	Moisture (%)	L*	a*	b*
1	31.77	3.54	76.55	2.45	75.44
1	29.31	3.47	76.54	2.46	75.42
1	30.23	3.49	76.54	2.49	75.49
2	47.44	4.45	84.26	−2.90	40.89
2	46.74	4.50	84.27	−2.86	40.89
2	47.72	4.39	84.26	−2.85	40.92
3	29.53	4.28	78.74	5.75	72.66
3	29.77	4.25	78.73	5.73	72.61
3	30.89	4.30	78.74	5.76	72.65
4	33.82	3.42	72.69	9.00	76.82
4	32.77	3.32	72.76	8.99	76.86
4	33.63	3.38	72.91	8.92	76.81
5	17.20	3.40	76.34	5.05	69.39
5	17.09	3.30	76.20	5.05	69.34
5	17.23	3.34	76.47	5.02	69.29
6	28.70	3.51	83.43	−2.59	54.58
6	28.91	3.47	83.42	−2.59	54.50
6	28.99	3.54	83.43	−2.62	54.53
7	34.19	3.35	86.49	−6.10	47.74
7	34.33	3.30	86.49	−6.10	47.80
7	34.26	3.35	86.52	−6.10	47.81

```
Brand = as.factor (Example6_6 $ Brand)
Y <- cbind (Example6_6 $ Fat.content, Example6_6 $ moisture, Example6_6 $ L, Example6_6 $ a, Example6_6 $ b)
Ma <- manova (Y ~ Brand)
summary (Ma, "Wilks")
summary (Ma, "Hotelling-Lawley")
summary (Ma, "Pillai")
summary (Ma, "Roy")
```

The results of using R commands to analyze the data regarding fish crackers, including the four multivariate tests and the p-value, are presented below.

```
> Brand = as.factor (Example6_6 $ Brand)
> Y <- cbind (Example6_6 $ Fat.content, Example6_6 $ moisture, Example6_6 $ L, Example6_6 $ a, Example6_6 $ b)
> Ma <- manova (Y ~ Brand)
> summary (Ma, "Wilks")
                Df      Wilks    approx F   num Df   den Df      Pr(>F)
Brand            6  3.0095e-17     18900       30       42    < 2.2e-16 ***
Residuals       14
---
Signif. codes: 0 '***' 0.001 '**' 0.01 '*' 0.05 '.' 0.1 ' ' 1

> summary (Ma, "Hotelling-Lawley")
                Df  Hotelling-Lawley   approx F   num Df   den Df      Pr(>F)
Brand            6            250900      70252       30       42    < 2.2e-16 ***
Residuals       14
---
Signif. codes: 0 '***' 0.001 '**' 0.01 '*' 0.05 '.' 0.1 ' ' 1

> summary (Ma, "Pillai")
                Df    Pillai   approx F   num Df   den Df   Pr(>F)
Brand            6   4.9745     456.08       30       70    < 2.2e-16 ***
Residuals       14
---
Signif. codes: 0 '***' 0.001 '**' 0.01 '*' 0.05 '.' 0.1 ' ' 1

> summary (Ma, "Roy")
                Df      Roy    approx F   num Df   den Df   Pr(>F)
Brand            6   224925     524825        6       14    < 2.2e-16 ***
Residuals 14
---
Signif. codes: 0 '***' 0.001 '**' 0.01 '*' 0.05 '.' 0.1 ' ' 1
```

Based on the MANOVA results for the selected parameters in fish crackers, we have determined that the brands are different since the p-values for all of the multivariate tests are very small, as presented in the output (2.2e-16). The brand of fish crackers has a significant effect on at least one response (parameter) measured from the fish cracker samples. This result could be due to the difference in the technology used to produce the crackers, the type of packaging, the material used, or other reasons.

6.4.2   Two-Way MANOVA

Suppose there are two factors: factor A has **a** levels and factor B has **b** levels. The objective is to investigate the effect of different levels of each factor on several responses measured on the same experimental unit. In a two-way multivariate analysis of variance (MANOVA), the researchers are interested in three hypotheses—the first two hypotheses are related to the main effect of each factor and the third hypothesis is related to the interaction between the different levels of factor A and factor B. The entries for a two-way MANOVA are given in Table 6.14.

The hypotheses for interaction and main effects can be tested by using the Wilk's test, as shown in (6.3) and (6.4).

$$\text{For the main effect, } = \frac{|E|}{|E + SS_A|}, \Lambda_B = \frac{|E|}{|E + SS_B|} \tag{6.3}$$

and for the interaction,

$$\Lambda_{AB} = \frac{|E|}{|E + SS_{AB}|} \tag{6.4}$$

Example 6.7: Removal of toxic heavy metals—We investigate the effects of the following four pectin samples: modified durian rind pectin (mDRP), durian rind pectin (DRP), citrus pectin (CP), and modified citrus pectin (mCP). In

TABLE 6.14 The Entries of a Two-Way MANOVA

Source of variation (S.O.V)	Degrees of freedom (d.f)	Sum of squares and products matrices (SSP)
Factor (A)	$a-1$	$SS_A = \sum_{i=1}^{a} bn(\overline{X}_{i.} - \overline{X})(\overline{X}_{i.} - \overline{X})'$
Factor (B)	$b-1$	$SS_B = \sum_{j=1}^{b} kn(\overline{X}_{.j} - \overline{X})(\overline{X}_{.j} - \overline{X})'$
Interaction (AB)	$(a-1)(b-1)$	$SS_{AB} = \sum_{i=1}^{a}\sum_{j=1}^{b} n(\overline{X}_{ij} - \overline{X}_{i.} - \overline{X}_{.j} + \overline{X})(\overline{X}_{ij} - \overline{X}_{i.} - \overline{X}_{.j} + \overline{X})'$
Residual (Error)	$ab(n-1)$	$E = SS_{Error} = \sum_{i=1}^{a}\sum_{j=1}^{b}\sum_{k=1}^{n}(\overline{X}_{ijk} - \overline{X}_{ij})(\overline{X}_{ijk} - \overline{X}_{ij})'$
Total	$abn-1$	$E = SS_{Error} = \sum_{i=1}^{a}\sum_{j=1}^{b}\sum_{k=1}^{n}(\overline{X}_{ijk} - \overline{X})(\overline{X}_{ijk} - \overline{X})'$

addition, we investigate four different concentrations of biosorbent (0.1, 0.2, 0.5, and 1) on the percentage removal of toxic heavy metals. We must test whether the different pectin samples and the concentrations of biosorbent have a significant effect on the removal of toxic heavy metals. Five heavy metals were measured from different samples, namely, Cd, Pb, Zn, Cu, and Ni. Three replicates were used for each combination of pectin samples, and the concentration of biosorbent. Assume the data are normally distributed. The data are given in Table 6.15. The data had been stored as a .CSV (Example6_7) file.

Two variables (independent variables or factors), pectin samples and concentrations of biosorbent, were thought to be influential factors on the percentage removal of toxic heavy metals. The data for the heavy metals' removal were analyzed using two-way analysis of variance (MANOVA) to ascertain the effect of different pectin samples and different concentrations of biosorbent on the percent removal of heavy metal ions. Furthermore, the effects of the interactions between the two selected factors on the percent removal of toxic heavy metals were investigated.

The function manova()was also used to perform a two-way multivariate analysis of variance (MANOVA). The data should be prepared and combined in a file similar to one-way MANOVA by using the function cbind() before applying the command manova(). The last function is summary(), which provides detailed results regarding the four multivariate tests.The results of using built-in functions and commands in R to perform a two-way MANOVA for the percent removal of toxic heavy metals are shown below.

```
#Define the responses
Cd = Example6_7 $ Cd
Pb = Example6_7 $ Pb
Zn = Example6_7 $ Zn
Cu = Example6_7 $ Cu
Ni = Example6_7 $ Ni
Y <- cbind (Cd, Pb, Zn, Cu, Ni)
Type = as.factor (Example6_7 $ Type)
Con. = as.factor (Example6_7 $ Con.)
Ma <- manova (Y ~ Type*Con.)
summary (Ma, test = "Wilks")
summary (Ma, test = "Hotelling-Lawley")
summary (Ma, test = "Pillai")
summary (Ma, test = "Roy")
```

TABLE 6.15 Percent Removal of Heavy Metal Ions by Different Pectins

Type	Con.	Cd(II)	Cu(II)	Zn(II)	Pb(II)	Ni(II)
CP	0.10	8.90	25.93	6.18	7.02	8.20
CP	0.20	14.38	29.63	10.67	18.42	14.75
CP	0.50	19.86	55.56	29.21	39.47	31.15
CP	1.00	39.04	81.48	39.89	66.67	49.18
CP	0.10	8.16	23.08	5.88	8.04	7.14
CP	0.20	12.24	34.62	11.18	17.86	14.29
CP	0.50	21.77	50.00	28.82	41.07	37.50
CP	1.00	40.82	80.77	40.59	68.75	53.57
CP	0.10	7.64	22.22	6.98	7.21	9.09
CP	0.20	13.89	29.63	9.30	16.22	12.73
CP	0.50	21.53	59.26	25.58	36.94	32.73
CP	1.00	35.42	77.78	44.19	65.77	54.55
mCP	0.10	17.81	22.22	14.61	14.04	9.84
mCP	0.20	28.08	48.15	17.42	23.68	24.59
mCP	0.50	61.64	81.48	65.73	68.42	54.10
mCP	1.00	93.42	98.59	95.51	98.07	92.95
mCP	0.10	18.37	19.23	14.12	14.29	8.93
mCP	0.20	25.17	42.31	17.65	23.21	26.79
mCP	0.50	65.99	84.62	61.18	70.54	57.14
mCP	1.00	94.15	98.31	94.12	97.77	92.86
mCP	0.10	15.97	18.52	13.37	13.51	9.09
mCP	0.20	27.08	40.74	19.77	21.62	21.82
mCP	0.50	68.06	81.48	56.98	70.27	60.00
mCP	1.00	94.31	98.30	93.60	97.84	92.36
DRP	0.10	3.42	7.41	3.37	5.26	6.56
DRP	0.20	5.48	18.52	5.62	8.77	14.75
DRP	0.50	8.22	33.33	7.87	23.68	22.95
DRP	1.00	10.96	59.26	8.43	39.47	27.87
DRP	0.10	4.08	7.69	2.94	4.46	7.14
DRP	0.20	6.12	15.38	5.29	8.04	14.29
DRP	0.50	9.52	34.62	7.65	21.43	19.64
DRP	1.00	10.88	50.00	7.65	37.50	23.21
DRP	0.10	3.47	7.41	3.49	4.50	7.27
DRP	0.20	6.25	18.52	4.65	8.11	14.55
DRP	0.50	9.72	33.33	8.72	21.62	20.00
DRP	1.00	9.72	55.56	9.30	38.74	27.27

(Continued)

TABLE 6.15 (Continued)

Type	Con.	Cd(II)	Cu(II)	Zn(II)	Pb(II)	Ni(II)
mDRP	0.10	8.22	25.93	5.06	1.75	11.48
mDRP	0.20	10.27	37.04	9.55	11.40	22.95
mDRP	0.50	35.62	62.96	19.10	28.95	34.43
mDRP	1.00	41.10	81.48	42.13	58.77	42.62
mDRP	0.10	9.52	26.92	5.88	1.79	12.50
mDRP	0.20	12.24	30.77	10.59	11.61	19.64
mDRP	0.50	34.01	53.85	15.88	32.14	35.71
mDRP	1.00	40.82	80.77	40.59	57.14	42.86
mDRP	0.10	9.03	22.22	5.81	1.80	10.91
mDRP	0.20	11.81	33.33	11.63	12.61	21.82
mDRP	0.50	31.25	59.26	18.02	27.93	30.91
mDRP	1.00	38.19	81.48	44.77	57.66	43.64

The first five rows were used to define and extract the selected responses from the file `Example6_7`. Then we combined all of the selected heavy metals (the dependent variables and responses) in one file using the `cbind()` function and placed the result in a new location called `Y`. The independent variable (factor) was defined to be the `Type` and the concentration (`con.`) as factors. The entries for two-way MANOVA for the percent removal of toxic heavy metals with the p-values for the four multivariate tests are given below.

```
> Cd = Example6_7 $ Cd
> Pb = Example6_7 $ Pb
> Zn = Example6_7 $ Zn
> Cu = Example6_7 $ Cu
> Ni = Example6_7 $ Ni
> Y <- cbind (Cd, Pb, Zn, Cu, Ni)
> Type = as.factor (Example6_7 $ Type)
> Con. = as.factor (Example6_7 $ Con.)
> Ma <- manova (Y ~ Type*Con.)
> summary (Ma, test = "Wilks")
           Df       Wilks  approx F  num Df  den Df   Pr(>F)
Type        3  1.3642e-05   294.519      15  77.697  < 2.2e-16 ***
Con.        3  1.5442e-04   119.254      15  77.697  < 2.2e-16 ***
Type:Con.   9  6.1440e-06    38.858      45 128.354  < 2.2e-16 ***
Residuals  32
---
Signif. codes: 0 '***' 0.001 '**' 0.01 '*' 0.05 '.' 0.1 ' ' 1
> summary (Ma, test = "Hotelling-Lawley")

          Df Hotelling-Lawley  approx F  num Df  den Df   Pr(>F)
Type       3           832.18   1479.44      15      80  < 2.2e-16 ***
Con.       3          1317.49   2342.20      15      80  < 2.2e-16 ***
Type:Con.  9           410.00    240.54      45     132  < 2.2e-16 ***
Residuals 32
---
Signif. codes: 0 '***' 0.001 '**' 0.01 '*' 0.05 '.' 0.1 ' ' 1
```

(Continued)

(Continued)

```
> summary (Ma, test = "Pillai")
          Df  Pillai  approx F  num Df  den Df      Pr( > F)
Type       3  2.7367   62.374      15      90   < 2.2e-16 ***
Con.       3  2.0619   13.187      15      90   < 2.2e-16 ***
Type:Con.  9  3.8244   11.566      45     160   < 2.2e-16 ***
Residuals 32
---
Signif. codes: 0 '***' 0.001 '**' 0.01 '*' 0.05 '.' 0.1 ' ' 1

> summary (Ma, test = "Roy")
          Df     Roy  approx F  num Df  den Df      Pr( > F)
Type       3  810.51  4863.1        5      30   < 2.2e-16 ***
Con.       3 1314.87  7889.2        5      30   < 2.2e-16 ***
Type:Con.  9  388.29  1380.6        9      32   < 2.2e-16 ***
Residuals 32
---
Signif. codes: 0 '***' 0.001 '**' 0.01 '*' 0.05 '.' 0.1 ' ' 1
```

The interpretation of the results should start with the interactions between different pectin samples and different concentrations of biosorbent. These interactions exhibited a significant effect on the removal of toxic heavy metals, as the p-value is very small (p-value $<2.2e\text{-}16$) for all multivariate tests. The significant interaction indicates that the different pectin samples and concentrations of biosorbent depend on each other during the removal of toxic heavy metals. Furthermore, the results of the analysis showed that different pectin samples and concentrations of biosorbent affect significantly the behavior of at least one response (p-value $<2.2e\text{-}16$), and as a result, these factors will affect the percentage removal of heavy metals' ions.

Example 6.8: Effects of date palm fruit crude on lipid oxidation—A researcher wants to investigate the effects of date palm fruit crude aqueous extract (DPF-CE) on the lipid oxidation of minced chicken meat (MCM) during chilled storage. The effects of two factors on the three responses of the pH, the peroxide value (PV), and the thiobarbituric acid (TBA) of MCM were studied; the two factors are the types of dates (Bam and Kharak) and the concentrations of extract (0, 2, and 4). Three replicates at each combination of types of dates and each concentration of extract were used. Assume that the data are normally distributed. The data are given in Table 6.16; they had been stored as a `.csv` (`Example6_8`) file.

The effects of the types of dates and the concentration of the extract on lipid oxidation were investigated through the behavior of these three responses: the pH, PV, and TBA. The data for the lipid oxidation were analyzed using built-in functions in R to conduct MANOVA. The command used to perform MANOVA for the data in Table 6.16 is similar to the commands used in Example 6.7 if we account for the fact that the entries are for lipid oxidation data.

The results of using R commands to analyze the effects of the types of dates and the concentrations of the extract on the pH, PV, and TBA of MCM are given below.

```
> pH = Example6_8 $ pH
> PV = Example6_8 $ PV
> TBA = Example6_8 $ TBA
> Y <- cbind (pH, PV, TBA)
> Types_dates = as.factor (Example6_8 $ Typesofdates)
> concentration = as.factor (Example6_8 $ concentration)
> Ma <- manova (Y ~ Types_dates*concentration)
> summary (Ma, test = "Pillai")
```

(Continued)

(Continued)

```
                 Df    Pillai   approx F   num Df   den Df      Pr(>F)
Types_dates       1   0.98035   166.322       3       10    7.861e-09 ***
concentration     2   1.93269   105.279       6       22    4.598e-15 ***
Types_dates:
concentration     2   1.08099     4.313       6       22    0.005057 **
Residuals        12
---
Signif. codes: 0 '***' 0.001 '**' 0.01 '*' 0.05 '.' 0.1 ' ' 1
> summary (Ma, test = "Hotelling-Lawley")

                 Df   Hotelling-Lawley   approx F   num Df   den Df      Pr(>F)
Types_dates       1             49.90     166.32        3       10    7.861e-09 ***
concentration     2            479.02     718.53        6       18    < 2.2e-16 ***
Types_dates:
concentration     2            102.44     153.65        6       18    1.840e-14 ***
Residuals        12
---
Signif. codes: 0 '***' 0.001 '**' 0.01 '*' 0.05 '.' 0.1 ' ' 1
> summary (Ma, test = "Roy")
                 Df     Roy   approx F   num Df   den Df      Pr(>F)
Types_dates       1   49.90    166.32       3       10    7.861e-09 ***
concentration     2  464.67   1703.81       3       11    5.967e-15 ***
Types_dates:
concentration     2  102.34    375.23       3       11    2.346e-11 ***
Residuals        12
---
Signif. codes: 0 '***' 0.001 '**' 0.01 '*' 0.05 '.' 0.1 ' ' 1
> summary (Ma, test = "Wilks")

                 Df     Wilks      approx F   num Df   den Df      Pr(>F)
Types_dates       1  0.0196477    166.322       3       10    7.861e-09 ***
concentration     2  0.0001399    278.450       6       20    < 2.2e-16 ***
Types_dates:
concentration     2  0.0087997     32.201       6       20    2.914e-09 ***
Residuals 12
Signif. codes: 0 '***' 0.001 '**' 0.01 '*' 0.05 '.' 0.1 ' ' 1
```

TABLE 6.16 The Results of the Experimental Design for Two Factors That Affect the pH, PV, and TBA

Types of dates	Concentration	pH	PV	TBA
Bam	0	6.92	16.30	4.54
Bam	0	7.11	15.30	4.47
Bam	0	7.50	15.80	4.52
Bam	2	6.59	11.10	1.88
Bam	2	6.45	12.50	1.61

(Continued)

TABLE 6.16 (Continued)

Types of dates	Concentration	pH	PV	TBA
Bam	2	6.52	11.70	1.75
Bam	4	4.79	11.20	3.27
Bam	4	4.82	9.20	3.24
Bam	4	4.80	10.20	3.26
Kharak	0	7.16	15.80	4.33
Kharak	0	7.20	16.20	4.45
Kharak	0	7.18	16.10	4.38
Kharak	2	6.72	13.20	1.89
Kharak	2	5.92	12.40	1.76
Kharak	2	6.45	12.80	1.83
Kharak	4	5.12	10.20	0.80
Kharak	4	5.10	9.40	0.97
Kharak	4	5.70	9.80	0.89

The selected responses were extracted from the file `Example6_8`. Then a two-way MANOVA was carried out. The MANOVA results showed that there were interactions between the types of dates and the concentrations of extract, as the p-value is very small; it is presented in the `Pr(>F)` column. This finding indicates that the two factors work together to influence the selected responses of the pH, PV, and TBA. Furthermore, both the types of dates and the concentrations of extract had a significant effect on at least one of the responses. Thus the types of dates and the concentrations of extract should be considered influential factors for any further study.

FURTHER READING

Alkarkhi, A. F. M., & Low, H. C. (2012). *Elementary statistics for technologist*. Pulau Pinang, Malaysia: Universiti Sains Malaysia press.

Bryan, F. J. M. (1991). *Multivariate statistical methods : A primer*. Great Britain: Chapman&Hall.

Daniel, H. (2013). *Blog Archives, High Resolution Figures in R* [Online]. Available: <https://www.r-bloggers.com/author/daniel-hocking/> Accessed 15.07.17.

Biglari, F., AlKarkhi, A. F. M., & Easa, A. M. (2008). Antioxidant activity and phenolic content of various date palm (*Phoenix dactylifera*) fruits from Iran. *Food Chemistry, 107*, 1636−1641.

Johnson, R. A., & Wichern, D. W. (2002). *Applied multivariate statistical analysis*. New Jersey: Prentice Hall.

Kabacoff, R. I. (2014). *Quick-R, accessing the power of R* [Online]. Available: <http://www.statmethods.net/stats/anova.html> Accessed 09.09.16.

Mark, G. *Using R for statistical analyses - ANOVA* [Online]. Available: <http://www.gardenersown.co.uk/education/lectures/r/anova.htm> Accessed 07.07.16].

Rencher, A. C. (2002). *Methods of multivariate analysis*. New York: J. Wiley.

Saifullah, R., Alkarkhi, A. F. M., Yeoh, S. Y., Liong, M.-T., & Azhar, M. E. (2009). Effect of banana pulp and peel flour on physicochemical properties and in vitro starch digestibility of yellow alkaline noodles. *International Journal of Food Sciences and Nutrition, 60*(S4), 326340.

Sarah, S. (2012). *Instant R-Creating an interaction plot in R* [Online]. WordPress | Nest Theme by YChong. Available: <http://www.instantr.com/2012/12/13/creating-an-interaction-plot-in-r/>.

Shin, Y. Y., Alkarkhi, A. F. M., Saifullah, B. R., & Azhar, M. E. (2011). Effect of cooking on physical and sensory properties of fresh yellow alkaline noodles prepared by partial substitution of wheat flour with soy protein isolate and treated with cross-linking agents. *International Journal of Food Sciences and Nutrition, 62*, 410−417.

Wong, W. W., Alkarkhi, A. F. M., & Azhar, M. E. (2010). Comparing biosorbent ability of modified citrus and durian rind pectin. *Carbohydrate Polymers, 79*, 584−589.

Chapter 7

Regression Models

Learning outcomes

At the end of this chapter, you should be able

- *To describe simple, multiple, and multiple multivariate linear regression models.*
- *To know how to use analysis of variance (ANOVA) in regression.*
- *To understand hypothesis testing in multiple linear regression.*
- *To know how to test the overall significance of the regression model.*
- *To know tests of the individual regression coefficients regression model.*
- *To understand tests on a subset of the regression coefficients.*
- *To know how to interpret regression models.*
- *To explain the coefficient of determination.*
- *To apply R commands to perform linear regression models.*
- *To report on the useful conclusions.*

7.1 INTRODUCTION

Regression analysis is a statistical technique used to describe the relationship between two sets of variables—one set includes one or more variables, called the independent or explanatory variables, which affect the second set of variables, which includes one or more variables called the dependent (response) variables. Regression analysis is also used to predict the values of one or more response variables with the equation of a straight line that describes the relationship between the two sets of variables. The straight line is called a regression line, and the equation is known as a regression equation.

In this chapter, three cases of regression models will be discussed based on the number of variables in the model as given below.

1. Simple linear regression: Simple linear regression is used to describe the relationship between one dependent (response) variable Y and one independent (explanatory)variable X. For example, the study of the effect of FRAP as an independent variable on the total phenolic content (TPC) (mg GAE/100 g dw) as a dependent variable.
2. Multiple linear regression: Multiple linear regression is used to describe the relationship between one dependent variable Y and several independent variables $X_1, X_2, \ldots, X_k$. For example, we study the effects of a date palm fruit's crude aqueous extract (DPF-CE) on the lipid oxidation of minced chicken meat (MCM) during chilled storage. The three factors we studied were the types of dates (Bam and Kharak); the concentration of extract; and the storage time of the pH, peroxide value (PV), and thiobarbituric acid (TBA) of MCM.
3. Multivariate multiple linear regression: Multivariate multiple regression consists of multiple dependent variables and any number of independent variables. For example, a researcher wants to investigate the effect of eating habits, such as how the number of ounces of chocolate, fish, and dairy products consumed per day affects a person's cholesterol, blood pressure, and weight.

7.2 REGRESSION ANALYSIS IN R

The R statistical software offers several built-in functions to perform regression analysis. These functions cover fitting a regression model, hypothesis testing for the model (the ANOVA Table and t-test), predicting new values, and calculating fitted values and residuals. The following functions are used in R to conduct regression analysis.

Easy Statistics for Food Science with R. DOI: https://doi.org/10.1016/B978-0-12-814262-2.00007-8

1. Either the function `lm()` or the function `Coef(lm())` can be used to fit a regression model (intercept and slope).

```
lm(y ~ x, data = data frame)
or
Coef (lm(y ~ x, data = data frame))
```

where y represents the dependent variable and x represents the independent variable.

1. The `summary()` function is used to provide detailed information on the estimated coefficients, residuals, t-test, p-value, coefficient of determination (R^2), and estimated error.

```
Summary (data frame)
```

2. The function `anova()` is used to produce an ANOVA Table, including the degrees of freedom (`Df`), sum of squares (`Sum sq`), mean sum of squares (`Mean sq`), F value, and p-value (`pr(>F)`).

```
anova (response ~ independent variables, data = data frame)
or
anova (lm(y ~ x, data = data frame))
```

3. In the case of k independent variables (explanatory), the same functions `lm()` or `Coef(lm())` can be used to fit a multiple linear regression model.

```
lm (y ~ x1 + x2 + ...+ Xk, data = data frame)
or
coef (lm(y ~ x1 + x2 + ...+ Xk, data = data frame))
```

where y represents the dependent variable (response) and x1, x2, ..., xk are independent variables.

4. The `predict()` function is used to make a prediction. This function provides the predicted value with lower and upper confidence limits. Three R commands are used for prediction.

```
Model = lm(y ~ x, data = data frame)
Newvalue = data.frame (x = give value)
Predict (model, newvalue, interval = "predict")
```

5. The function `fitted()` is used to calculate the fitted values.

```
fitted (data frame)
```

6. The function `resid()` is used to produce the residuals.

```
resid (data frame)
```

7.3 SIMPLE LINEAR REGRESSION

Simple linear regression studies the relationship between one dependent variable Y and one independent variable X. The simple linear regression equation is given in (7.1),

$$Y = \beta_0 + \beta_1 X + \varepsilon \tag{7.1}$$

where β_0 represents the intercept of the regression equation, β_1 is the slope of the regression equation, and ε is the error term. β_0 and β_1 are parameters to be estimated from sample data. The estimates of β_0 and β_1 are b_0 and b_1, respectively. The regression equation obtained after estimation of b_0 and b_1 is called the predicted, estimated, or fitted model. The fitted model is given in (7.2),

$$\hat{Y} = b_0 + b_1 X \tag{7.2}$$

where $\hat{Y}$ represents the fitted response. b_0 and b_1 are calculated using the least squares method; the formulas are given in (7.3) and (7.4), respectively.

$$b_1 = \frac{n \sum_{i=1}^{n} X_i Y_i - \sum_{i=1}^{n} X_i \sum_{i=1}^{n} Y_i}{n \sum_{i=1}^{n} X_i^2 - [\sum_{i=1}^{n} X_i]^2} \tag{7.3}$$

$$b_0 = \overline{Y} - b_1 \overline{X} \tag{7.4}$$

Note
- The line of best fit is obtained by graphing the regression equation.
- The dependent variable Y is normally distributed.
- The sign of the slope of the regression line b_1 shows the direction of the relationship between the dependent and the independent variable, whether it is positive (increase) or negative (decrease).

7.3.1 Hypothesis Testing

Hypothesis testing regarding simple linear regression tests the effect of the independent variable on the model, determining whether or not the selected variable contributes to explaining the total variation in the data. This test can be carried out by testing the slope β_1 of the regression model. The null and alternative hypotheses are given below.

$$H_0\colon \beta_1 = \beta_{1,0}$$

$$H_1\colon \beta_1 \neq \beta_{1,0}$$

The test statistic used for this test is

$$t = \frac{b_1 - \beta_{1,0}}{se(b_1)} \text{ with } n - 2 \text{ degrees of freedom.}$$

where b_1 is the least square estimate of β_1, $\beta_{1,0}$ is a constant (given value), and $se(b_1)$ is the standard error. The value of $se(b_1)$ can be calculated as given in (7.5).

$$se(\hat{\beta}_1) = \sqrt{\frac{\sum_{i=1}^{n} e_i^2 / (n-2)}{\sum_{i=1}^{n} (X_i - \overline{X})^2}} \tag{7.5}$$

The null hypothesis states that there is no effect of the independent variable on the response (output) variable, while the alternative hypothesis states that the selected independent variable has an effect on the response and should be kept in the model. Rejection of the null hypothesis indicates that the selected independent variable contributes to explaining the variation in the observed values of the dependent variable. In other words, the relationship between the independent variable and dependent variable exists, and the model can be used to describe the behavior of the process. Failure to reject the null hypothesis indicates that no linear relationship exists between the independent variable and dependent variable, which means that the independent variable does not contribute to explaining the variation in the observed values of the dependent variable.

A similar procedure can be used to test the hypothesis on the intercept β_0, to determine whether or not the intercept should be kept in the model. The null and alternative hypotheses are:

$$H_0: \beta_0 = \beta_{0,0}$$

$$H_1: \beta_0 \neq \beta_{0,0}$$

moreover, the test statistic used is:

$$t = \frac{b_0 - \beta_{0,0}}{se(b_0)}$$

where b_0 is the least squares estimate of β_0 and $se(b_0)$ is its standard error. The value of $se(b_0)$ can be calculated as given in (7.6).

$$se(b_0) = \sqrt{\frac{\sum_{i=1}^{n} e_i^2}{n-2}\left[\frac{1}{n} + \frac{\overline{X}^2}{\sum_{i=1}^{n}(X_i - \overline{X})^2}\right]} \tag{7.6}$$

Note
- The test regarding the slope and intercept can be carried out using a t test or an analysis of variance (ANOVA) approach to test the significance of the regression.

7.3.2   Interpretation of Regression Equation

Interpreting the regression equation depends on the regression coefficient associated with the independent variable as well as the sign. The regression coefficient gives the relative contribution of each independent variable to the response; this contribution can be measured directly by the regression coefficient in the fitted model, and the sign gives the direction of the influence on the dependent variable, positive or negative.

If the regression coefficient in the fitted model is positive, the independent variable has the ability to increase the response. If the coefficient is negative, the independent variable can decrease the response.

Example 7.1: Antioxidants in dates—A researcher wants to analyze the edible parts of date palm fruits (DPF) for their antioxidant activities using the ferric reducing/antioxidant method (FRAP) (μ m 01/100 g dw). Describe the effect of the FRAP on the total phenolic content (TPC) (mg GAE/100 g dw) in dates. Use the following data to fit a regression model for FRAP and TPC. The data for the FRAP and the total phenolic content in dates are given in Table 7.1.

The regression equation that describes the relationship between the total phenolic content in dates (TPC) as a response variable (dependent) and the FRAP as an independent variable can be calculated by applying formulas (7.3) and (7.4) to obtain the coefficients of the regression equation. Built-in functions in R can be used to calculate the coefficients of the regression model and provide detailed information regarding its significance. The results can enable researchers to make a decision regarding the model using features such as the ANOVA Table, p-value, and t-test.

R provides two equivalent functions, `lm()` and `coef(lm(y ~ x))`, that can be used to calculate the regression coefficients. The data had been stored as a `.CSV` (`Example7_1`) file. The stored data can be imported into an R environment. Then one can apply R's built-in functions to calculate the coefficients of the regression model that describes the relationship between TPC (y) and FRAP (x).

TABLE 7.1 The Data for the FRAP and the Total Phenolic Content in Dates

FRAP = X	20.00	26.93	16.00	13.32	29.34	11.66	19.12
TPC = Y	2.71	4.80	2.23	1.60	4.40	2.19	3.23

```
> model = lm(y ~ x, data = Example7_1)
> model

Call:
lm(formula = y ~ x, data = Example7_1)

Coefficients:
(Intercept)          x
   -0.2655     0.1688
```

The first line of the results produced by R commands is a call (`lm(y ~ x, data = Example7_1)`) to fit a regression model to the available data, then a call to print the results (`> model`). The intercept coefficient (b_0) and the slope (b_1) for the regression model of antioxidants are -0.2655 and 0.1688, respectively. The next step is to write the model as given in (7.2).

$$\hat{Y} = -0.2655 + 0.1688X$$

Observe that the sign of b_1 in the fitted model is positive. This sign indicates that FRAP (X) affects positively the content of TPC (Y), which means that if the FRAP increases by one unit, the TPC will increase by as much as 0.1688.

The significance of the model should be tested to make a decision whether the FRAP should be retained in the model (based on whether it is an influential variable). Thus the model can be tested by producing an ANOVA Table.

The function `anova()` was used to produce an ANOVA Table with a p-value to make a decision regarding the significance of the model for the total phenolic content in dates.

```
> anova (lm (y ~ x, data = Example7_1))
Analysis of Variance Table
Response: y
          Df  Sum Sq  Mean Sq  F value    Pr(>F)
x          1  7.5313   7.5313   37.247   0.00171 **
Residuals  5  1.0110   0.2022
---
Signif. codes: 0 '***' 0.001 '**' 0.01 '*' 0.05 '.' 0.1 ' ' 1
```

The results of the ANOVA showed that the model is significant, as the p-value under `Pr(>>F)` is small (p-value < 0.00171), which means that the FRAP contributes to explaining the variation in the TPC. In other words, the TPC depends on the FRAP. This result is a general one, and individual testing of the coefficients should be undertaken.

The researchers need more information about the analysis to build a comprehensive picture of the project. The function `summary()` provides more information on the model, including the estimated coefficients (`Estimate`), standard error (`Std. Error`), t-test (`t value`), p-value (`pr(>|t|)` for each coefficient, `residual standard error`, R-squared (R^2), and `adjusted R-squared`.

```
> summary (lm(y ~ x, data = Example7_1))

Call:
lm(formula = y ~ x, data = Example7_1)

Residuals:
      1       2        3        4        5       6       7
 -0.4004  0.5199  -0.2052  -0.3828  -0.2869  0.4874  0.2682
```

(Continued)

(Continued)

```
Coefficients:
            Estimate Std. Error t value Pr( > |t|)
(Intercept)  -0.26553   0.56498  -0.470   0.65815
x             0.16880   0.02766   6.103   0.00171 **
---
Signif. codes: 0 '***' 0.001 '**' 0.01 '*' 0.05 '.' 0.1 ' ' 1

Residual standard error: 0.4497 on 5 degrees of freedom
Multiple R-squared: 0.8816,  Adjusted R-squared: 0.858
F-statistic: 37.25 on 1 and 5 DF, p-value: 0.00171
```

The results of applying the function `summary()` for the model of the TPC and FRAP showed detailed information on the estimated coefficients. The intercept term b_0 did not show a significant effect, as the p-value is large (0.65815). However, the effect of FRAP (X) was significant on the TPC (p-value <0.00171). Thus the model will not have an intercept, so the intercept should be excluded from the model. The final model that describes the relationship between the TPC and the FRAP should be written as shown below.

$$\hat{Y} = 0.1688X$$

Furthermore, the value of the coefficient of determination R^2 is 0.88 (R^2 will be given in subsection 7.3.7). This value indicates that 88% of the total variance is explained by only one independent variable: FRAP (X).

Example 7.2: Cadmium and copper content in cockles—We examine the concentrations of two heavy metals (mg/L), cadmium ($Y = Cd$) as a response and copper ($X = Cu$) as an independent variable. Use the data given in Table 7.2 to study the relationship between Cu and Cd.

The relationship between the Cd and Cu can be studied by fitting the regression model that describes the nature of the relationship between cadmium (Cd) and copper (Cu). The function `lm(Cd ~ Cu)` in R was used to calculate the regression coefficients for the model to describe the relationship between Cd and Cu. The data had been stored as a .CSV file (`Example7_2`).

```
> model = lm(Cd ~ Cu, data = Example7_2)
> model

Call:
lm(formula = Cd ~ Cu, data = Example7_2)

Coefficients:
(Intercept)        Cu
     0.7977    0.4297
```

TABLE 7.2 The Data for the Cadmium and Copper Content in Cockles

$Y = Cd$	0.82	0.86	0.88	0.92	0.92	0.89	0.88	0.89	0.90	0.89
$X = Cu$	0.20	0.16	0.24	0.20	0.21	0.20	0.20	0.18	0.18	0.20
$Y = Cd$	0.90	0.89	0.84	0.86	0.88	0.91	0.91	0.88	0.90	0.92
$X = Cu$	0.18	0.17	0.17	0.17	0.18	0.19	0.17	0.16	0.17	0.17
$Y = Cd$	0.85	0.88	0.89	0.84	0.88	0.84	0.88	0.88	0.88	0.82
$X = Cu$	0.17	0.21	0.22	0.19	0.20	0.15	0.18	0.23	0.19	0.19
$Y = Cd$	0.93	0.88	0.93	0.89	0.84	0.86	0.89	0.87	0.79	0.85
$X = Cu$	0.20	0.19	0.20	0.18	0.20	0.17	0.17	0.17	0.15	0.19

The output of the function `lm(Cd ~ Cu)` consists of two values: the first value is the intercept coefficient b_0 = 0.7977, and the second value is the slope coefficient b_1 = 0.4297. No information is given regarding whether these values are significant. Hence the fitted regression model should include both coefficients as follows:

$$\hat{Y} = 0.7977 + 0.4297X$$

Before testing the model, the fitted regression model tells us in general that Y = Cd and X = Cu are positively correlated and that an increase of 1 unit in the concentration of Cu will increase the concentration of Cd by as much as 0.4297.

More information regarding the contribution of the Cu (the independent variable) and the intercept can be obtained from the analysis of variance using the `anova(lm(Cd ~ Cu))`and `summary(lm(Cd ~ Cu))` functions to produce an ANOVA Table, a t-test for each coefficient, a p-value, and a coefficient of determination (R^2).

```
> anova (lm(Cd ~ Cu, data = Example7_2))
Analysis of Variance Table

Response: Cd
          Df    Sum Sq    Mean Sq  F value  Pr(>F)
Cu         1  0.002906  0.0029059   3.1378  0.08452 .
Residuals 38  0.035192  0.0009261
---
Signif. codes: 0 '***' 0.001 '**' 0.01 '*' 0.05 '.' 0.1 ' ' 1
> summary (lm(Cd ~ Cu, data = Example7_2))

Call:
lm(formula = Cd ~ Cu, data = Example7_2)

Residuals:
      Min        1Q      Median       3Q       Max
 -0.072173  -0.017604   0.000639  0.020658  0.049233

Coefficients:
            Estimate Std. Error t value Pr(>|t|)
(Intercept)  0.79772    0.04544  17.557   <2e-16 ***
Cu           0.42971    0.24258   1.771   0.0845 .
---
Signif. codes: 0 '***' 0.001 '**' 0.01 '*' 0.05 '.' 0.1 ' ' 1

Residual standard error: 0.03043 on 38 degrees of freedom
Multiple R-squared: 0.07628, Adjusted R-squared: 0.05197
F-statistic: 3.138 on 1 and 38 DF, p-value: 0.08452
```

The results of applying the `anova()` and `summary()` functions showed that the model is not significant, as the p-value is greater than 0.05 (`pr(>F)` = 0.08452). This finding indicates that there is no significant relationship between the Cd and Cu. Additionally, the concentration of each parameter is not related to whether the other parameter is high or low in cockles.

Furthermore, the intercept (b_0) exhibited a significant effect, as indicated by the `t-value` test (the p-value is very small (2e-16)). However, the Cu did not have a significant effect on the Cd at a p-value < 0.05. The value of `Multiple R-squared` (R^2) is 0.07628, which is very low and indicates that most of the variance in the Cd is unexplained by the model.

We conclude that this model is negligible and does not help us describe the relationship between the cadmium and copper.

7.3.3 Prediction Using a Regression Equation

One objective of the regression analysis is to predict the response value at specified value of X. The prediction can be conducted by substituting the X value into the regression line to obtain the predicted Y value. The prediction should be made only when there is a significant relationship between the dependent and independent variables, which means that the correlation coefficient between the two variables is significantly different from 0 (there is a linear correlation). When the correlation is insignificant, the best prediction of Y (dependent variable) is the mean of the data values of Y.

Note
• Do not make prediction beyond the available values (i.e., beyond the range of data collected).

Example 7.3: Prediction of total phenolic content—Predict the total phenolic content (TPC) in dates when FRAP is 27 by using the regression equation line found in Example 7.1.

The regression equation was found in Example 7.1 to be

$$\hat{Y} = 0.2655 + 0.1688X$$

The prediction of the total phenolic content in dates when the FRAP is 27 requires a number of steps. The first step is to fit the regression model using the function `lm()`, the second step is to create a new `data.frame` for the required valued of X, and the last step is to apply the function `predict()`. These steps can be written as R commands that will calculate the predicted value of the TPC when the FRAP is equal to 27.

```
> model = lm(y ~ x, data = Example7_1)
> newdata = data.frame (x = 27)
> predict (model, newdata, interval = "predict")
      fit      lwr      upr
1 4.291963 2.945587 5.638339
```

The results produced by using the function `predict()` consist of the fitted value (`fit`), the lower confidence level (`lwr = 2.945587`), and the upper confidence level (`upr = 5.638339`). The value of the TPC will be `4.291963` when the value of the FRAP is 27. If the results of the t-test for Example 7.2 are used, then the regression model is

$$\hat{Y} = 0.1688X$$

and the predicted value is

$$\hat{Y} = 0.1688[27] = 4.558$$

Comparing the two results (including and excluding the intercept term) demonstrates that the difference between the two results is very small; this pattern could be due to the nonsignificant effect of the intercept b_0.

Note
• Fitted values can be calculated in R by using the `fitted()` function.

```
fitted () # predicted values
```

7.3.4 Outliers and Influential Observations

An outlier is defined as a point lying far away from the other points. An investigation should be conducted to determine whether or not this value is influential. One way to check is to draw a scatter plot and properly verify whether this value is an influential point or not. Each outlier should be checked properly to make a decision about whether to include or exclude it in the final analysis based on the importance of the value. If it is a real value and there is a scientific explanation for observing such an outlier value, then this value should be kept in the final analysis. Otherwise, the value should be excluded. The effect of an influential point is to pull the regression line toward the point itself, as illustrated in Example 7.4.

Example 7.4: Outlier observation—Use the data given in Example 7.2 for Cd and Cu, except that the last observation is changed to Cu = 1 and Cd = 3. The data are given in Table 7.3.

It is better to find the regression equation and the scatter diagram before and after changing the value of the Cu to illustrate the effect of the outlier. The function `lm(Cd ~ Cu)` was used to fit the two regression models before and after

TABLE 7.3 The Data for the Cadmium and Copper Content in Cockles After Changing the Two Values (Highlighted)

Y = Cd	0.82	0.86	0.88	0.92	0.92	0.89	0.88	0.89	0.90	0.89
X = Cu	0.20	0.16	0.24	0.20	0.21	0.20	0.20	0.18	0.18	0.20
Y = Cd	0.90	0.89	0.84	0.86	0.88	0.91	0.91	0.88	0.90	0.92
X = Cu	0.18	0.17	0.17	0.17	0.18	0.19	0.17	0.16	0.17	0.17
Y = Cd	0.85	0.88	0.89	0.84	0.88	0.84	0.88	0.88	0.88	0.82
X = Cu	0.17	0.21	0.22	0.19	0.20	0.15	0.18	0.23	0.19	0.19
Y = Cd	0.93	0.88	0.93	0.89	0.84	0.86	0.89	0.87	0.79	3.00
X = Cu	0.20	0.19	0.20	0.18	0.20	0.17	0.17	0.17	0.15	1.00

we changed the values. The data had been stored as the .CSV files (Example7_2a) and Example7_4 for the cases before and after we changed the value, respectively. The two models were fitted using the R command lm().

```
> model1 = lm(Cd1 ~ Cu1, data = Example7_2a) #Before changing
> model1

Call:
lm(formula = Cd1 ~ Cu1, data = Example7_2a)

Coefficients:
(Intercept)        Cu1
     0.7977    0.4297

> model = lm(Cd ~ Cu, data = Example7_4) # After changing
> model

Call:
lm(formula = Cd ~ Cu, data = Example7_4)

Coefficients:
(Intercept)        Cu
     0.4038    2.5552
```

The two regression models are

$$\hat{Y} = 0.7977 + 0.4297X \text{ (Before)}$$

$$\hat{Y} = 0.4038 + 2.5552X \text{ (After)}$$

Scatter plots for the data before and after the value was changed were produced to show the effects of the extreme values pictorially; see Fig. 7.1A and B, respectively (see the appendix for the R commands). The two scatter diagrams with a regression line were produced using R commands as shown in Fig. 7.1.

Applying these commands to the data before and after changing the value will give Fig. 7.1A and B.

a. Before
b. After

Observe that the coefficients of the regression models from before and after the change are quite different. Fig. 7.1A and B show the effect of the point (Cu = 1, Cd = 3), which is an influential point, and how it pulled the regression line toward itself. This value will change the results and lead to an incorrect conclusion about the relationship between the variables, which will mislead the researcher. Thus researchers should make a decision to retain or exclude this value

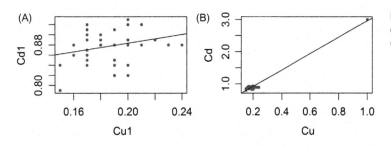

(A)

(B)

FIGURE 7.1   Scatter plot showing the effect of an outlier on regression equation (A) before we change the value and (B) after we change the value.

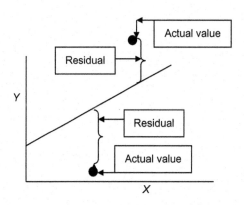

FIGURE 7.2   Showing the residual and the actual value.

based on scientific reasoning. The results of ANOVA after changing the value (not presented here) showed that the model is significant with a very small p-value.

7.3.5   Residuals

The residual is defined as the difference between the observed values of Y and the predicted values $\hat{Y}$ for a given X value. Residuals can be defined as the vertical distance between the observed value of Y and the regression line, as shown in Fig. 7.2.

Note
- The sum of the residuals for the entire data set is always zero.
- The sum of squares of the residuals is the smallest possible sum. The regression line is also called the least-squares line.

7.3.6   Explained and Unexplained Variation

The researchers are interested in identifying the sources of variation that affect the response (output) of the experiments. In other words, the researchers want to know the source of changes in the response variable Y and whether the changes are due to known or unknown sources to help understand the behavior of the experiment and make a decision.

In general, the total variation in the response Y is represented by $\sum (Y - \overline{Y})^2$, which can be divided into two parts:

- The first part is called explained variation due to the relationship between the independent variable X and the dependent variable Y. The explained variation is represented as $\sum (\hat{Y} - \overline{Y})^2$.
- The second part is called the unexplained variation due to an unknown source and is called the residual. Unexplained variation is represented as $\sum (Y - \hat{Y})^2$.

The total variation can be written as the sum of the two sources, explained and unexplained, as shown in (7.7).

$$\sum (Y - \overline{Y})^2 = \sum (\hat{Y} - \overline{Y})^2 + \sum (Y - \hat{Y})^2 \tag{7.7}$$

Example 7.5: Explained and unexplained variation—Use the data given in Example 7.1 to find the residuals, explained variation, unexplained variation, and total variation.

The fitted regression equation was calculated in Example 7.1 to be

$$\hat{Y} = -0.2655 + 0.1688X$$

The formula $Y - \hat{Y}$ is used to calculate the residual, where $\hat{Y}$ represents the predicted response (estimated response). $\hat{Y}$ can be calculated by substituting X values into the regression equation as shown below.

For the first X-value (20), the estimated value $\hat{Y}$ is

$$\hat{Y} = -0.2655 + 0.1688X = -0.2658 + 0.1688[20] = 3.11$$

The same operation will be repeated for the rest of X values to find the estimated values.

The total variation is computed by applying the formula $\sum (Y - \overline{Y})^2$. For the first value, the total variation is $(2.71 - 3.02)^2 = 0.008$. Continue the process for the other values and then find the sum of all values.

$$\sum (Y - \overline{Y})^2 = 8.54$$

The explained variation is computed by applying the formula $\sum (\hat{Y} - \overline{Y})^2$. For the first value, the explained variation is $(3.11 - 3.02)^2 = 0.01$. Continue the process for the other values.

$$\sum (\hat{Y} - \overline{Y})^2 = 7.53$$

The unexplained variation is computed by applying the formula $\sum (Y - \hat{Y})^2$. For the first value, the unexplained variation is $(2.71 - 3.11)^2 = 0.16$. Continue the process for the other values.

$$\sum (Y - \hat{Y})^2 = 1.01$$

Finally, it can be observed that the total variation is the sum of the explained and unexplained variations, as given in (7.7).

Total variation	=	Explained variation	+	Unexplained variation
8.54	=	7.53	+	1.01

The predicted values can easily be produced using the `predict()` function in R. Furthermore, the `resid()` built-in function is used to generate the residuals for the model. Explained and unexplained variations with the total variation are also calculated by R. The results of using R commands to calculate the predicted values and residual, as well as the explained, unexplained, and total variations, are presented below.

```
> model = lm(y ~ x, data = Example7_5)
> model

Call:
lm(formula = y ~ x, data = Example7_5)

Coefficients:
(Intercept)          x
   -0.2655     0.1688

> anova (lm(y ~ x, data = Example7_5))
Analysis of Variance Table

Response: y
          Df  Sum Sq  Mean Sq  F value  Pr(>F)
x          1  7.5313   7.5313   37.247  0.00171 **
Residuals  5  1.0110   0.2022
---
Signif. codes: 0 '***' 0.001 '**' 0.01 '*' 0.05 '.' 0.1 ' ' 1

> summary (lm(y ~ x, data = Example7_5))
```

(Continued)

(Continued)

```
Call:
lm(formula = y ~ x, data = Example7_5)

Residuals:
      1       2       3       4       5       6       7
-0.4004  0.5199 -0.2052 -0.3828 -0.2869  0.4874  0.2682

Coefficients:
            Estimate Std.  Error  t value  Pr(>|t|)
(Intercept) -0.26553    0.56498  -0.470   0.65815
x            0.16880    0.02766   6.103   0.00171 **
---
Signif. codes: 0 '***' 0.001 '**' 0.01 '*' 0.05 '.' 0.1 ' ' 1

Residual standard error: 0.4497 on 5 degrees of freedom
Multiple R-squared: 0.8816, Adjusted R-squared: 0.858
F-statistic: 37.25 on 1 and 5 DF, p-value: 0.00171
> predicted = round (predict(model), digits = 3)
> predicted
    1     2     3     4     5     6     7
3.110 4.280 2.435 1.983 4.687 1.703 2.962
> round (resid(model), digits = 3)
     1      2      3      4      5      6      7
-0.400  0.520 -0.205 -0.383 -0.287  0.487  0.268

> M = mean (Example7_5 $ y)
> round ((predicted - M)^2, digits = 3)
    1     2     3     4     5     6     7
0.008 1.580 0.346 1.081 2.769 1.742 0.004
> sum ((predicted - M)^2)              #Explained
[1] 7.529979
> round ((Example7_5 $ y - predicted)^2, digits = 3)
    1     2     3     4     5     6     7
0.160 0.270 0.042 0.147 0.082 0.237 0.072
> sum ((Example7_5 $ y - predicted)^2)    #Unexplained
[1] 1.010476
```

The results showed that the first step is to fit a regression model using the function `lm()`, and then to test the significance of the model by producing an ANOVA Table using the function `anova()` in R. The estimated values and residual values were calculated using the function `predict(model)` and the function `resid(model)`, respectively. The function `fitted()` can be used to calculate the estimated values instead of the function `predict()`. Furthermore, the mean for the dependent variable was calculated, as was the explained variation, and the last result was the sum of the unexplained variation (the sum of the squares of the errors). The `round()` function was used to round the results to three decimals to accommodate the entire result in a compact arrangement.

7.3.7 Coefficient of Determination

The coefficient of determination is a measure of the variation in Y that is explained by the regression line. The coefficient of determination is defined as the proportion of the variation in the dependent variable explained by the regression model (selected independent variable). The symbol for the coefficient of determination is R^2, and the formula for calculating R^2 is given in (7.8).

$$R^2 = \frac{\text{Explained variance}}{\text{Total variation}} \tag{7.8}$$

Note
- The result is usually expressed as a percentage.
- The value of R^2 lies between 0 and 1; $0 \leq R^2 \leq 1$.
- R^2 is a measure of the goodness of fit of the model.
- The square root of the coefficient of determination is called the correlation coefficient, which is a measure of the strength of the association between the two sets, the independent (explanatory) variable and the dependent variable.

Interpretation of R^2

R^2 is usually used to explain the changes of Y, which indicates the amount of the total variation in the dependent variable that is explained by the independent variable in the regression equation, while $1 - R^2$ represents the amount of variation in the dependent variable that is unknown (unexplained by the model).

Example 7.6: Coefficient of determination—Calculate the coefficient of variation (R-Squared or R^2) for the data presented in Example 7.5.

The value of the coefficient of variation is calculated based on the results of the explained and total variations in (7.8).

$$R^2 = \frac{\text{Explained variance}}{\text{Total variation}} = \frac{7.53}{8.54} = 0.88$$

Thus 88% of the total variation in the data for the TPC is explained by the model and can be accounted for by the independent variable (FRAP), and only 12% is due to unknown sources.

The summary() function in R can be used to produce R^2. The results of this function in Example 7.5 showed that the value of R^2 is Multiple R-squared: 0.8816 (at the bottom of the results of the summary() function with the F-statistics and p-value).

7.4 MULTIPLE REGRESSION

The relationship between one response variable and one independent variable was studied, and this relationship was modeled using simple linear regression. The estimated simple linear regression equation for one independent variable is:

$$\hat{Y} = b_0 + b_1 X$$

The idea of simple linear regression can be extended to include more than one independent variable $X_1, X_2, \ldots, X_k$ and one dependent variable Y. The equation that describes the relationship between the two sets of data (dependent and several independent variables) is called a multiple regression equation.

Multiple regression is used to examine the relationship between one dependent variable and several independent variables. The general form of a multiple regression equation is given in (7.9).

$$Y_i = \beta_0 + \beta_1 X_1 + \beta_2 X_2 + \ldots + \beta_k X_k + \varepsilon_i \tag{7.9}$$

Moreover, the fitted (estimated) equation is given in (7.10).

$$\hat{Y} = b_0 + b_1 X_1 + b_2 X_2 + \ldots + b_k X_k \tag{7.10}$$

where $\hat{Y}$ represents the fitted (predicted, estimated) value of the dependent variable; $\beta_1, \beta_2, \ldots, \beta_k$ are parameters to be estimated; and $X_1, X_2, \ldots, X_k$ are the independent variables. $b_0, b_1, \ldots, b_k$ are estimates of $\beta_1, \beta_2, \ldots, \beta_k$.

The least squares method is usually used to find the estimates of $\beta_1, \beta_2, \ldots, \beta_k$ using the formula given in (7.11) that minimizes the sum of squares of the deviation of the observed values Y from their fitted (predicted) values $\hat{Y}$,

$$b = (X'X)^{-1} X'Y \tag{7.11}$$

where $(X'X)$ is a nonsingular matrix. This condition holds if the number of observations n is at least one greater than the number of independent variables, $n > k + 1$.

Note
- The computations required for a multiple regression equation are complicated and require knowledge of advanced mathematics such as the inverse matrix and other operations related to matrices. However, for these computations, statistical packages can be used to analyze the data and to find the multiple regression equation. The important point is to focus on the interpretation of the results (output).

7.5 HYPOTHESIS TESTING IN MULTIPLE LINEAR REGRESSION

In multiple regression, there are three types of hypotheses tested on the regression coefficients $\beta's$. The three tests are as follows:

- Check the overall regression (test for the significance of the regression) using an ANOVA.
- Verify the importance of each independent variable in the model (regression coefficient) using the t-test.
- Test a subset of the regression coefficients. In other words, check the significance of some regression coefficients using a partial F test.

7.5.1 Test of Overall Significance of the Regression Model

The overall test of the model is a general test that gives information about whether or not the independent variables affect the response without providing information about each independent variable. This test is carried out using an analysis of variance (ANOVA).

The test is used to check if a linear relationship exists between the response variable and at least one of the independent (predictor) variables. The statements for the hypotheses are:

$$H_0: \beta_1 = \beta_2 = \ldots = \beta_k = 0$$

$$H_0 : \beta_i \neq 0 \text{ for at least one } i$$

The null hypothesis states that none of the selected variables affect the response (output), i.e., the effect of each independent variable is equal to zero, while the alternative hypothesis states that at least one independent variable affects the response. Rejecting the null hypothesis indicates that a relationship between the dependent variable and the set of independent variables exists, which means that the independent variables influence the observed value of the dependent variable. In contrast, failing to reject the hypothesis indicates that the relationship does not exist and that the model cannot be used to describe the behavior of the process.

7.5.2 Tests on Individual Regression Coefficients Regression Model

Tests on individual coefficient should be carried out if the overall test exhibited a significant effect of the selected independent variables. The t-test is used to check the significance of an individual coefficient to decide which variables affect the dependent variable. These variables are kept in the model, while insignificant variables should be excluded from the model. The statements for this test are:

$$H_0: \beta_i \neq 0$$

$$H_1: \beta_i \neq 0$$

The test statistic is:

$$t = \frac{b_i}{se(b_i)} \text{ with } n - 2 \text{ degrees of freedom.}$$

This test measures the contribution of each variable in explaining the variation in the data, while the remaining variables are included in the model.

7.5.3 Tests on a Subset of the Regression Coefficients

Tests on a subset of regression coefficients are used to test whether adding (deleting) variables to the model will increase (decrease) the regression sum of squares or not; in other words, testing the significance of adding (deleting) terms to the model to determine whether or not these terms contribute to explaining the variation in the dependent variable.

Assume that the vector of regression coefficients $\beta's$ is partitioned into two vectors. The first vector contains $k - r + 1$ coefficients, and the second contains r regression coefficients. The vector of the regression coefficient can be written as follows:

$$\beta = \begin{bmatrix} \beta_1 \\ \beta_2 \end{bmatrix}$$

where $\beta_1 = [\beta_0, \beta_1, \ldots, \beta_{k-r}]$ and $\beta_2 = [\beta_{k-r+1}, \beta_{k-r+2}, \ldots, \beta_k]$

The null and alternative hypotheses are:

$$H_0: \beta_2 = 0$$

$$H_1: \beta_2 \neq 0.$$

The test statistic used to check the contribution of the variables in β_2 is called the F test, as shown below.

$$F = \frac{SS_R(\beta_2 | \beta_1)/r}{MSE}$$

where $SS_R = (\beta_2 | \beta_1)$ represents the contribution of the variables in β_2. In other words, it represents the increase in the regression sum of squares when the variables corresponding to the coefficients in β_2 are added to a model already containing β_1. Rejecting the null hypothesis leads to the conclusion that at least one of the variables in β_2 contributes significantly to the regression model.

7.6   ADJUSTED COEFFICIENT OF DETERMINATION

The coefficient of determination R^2 has serious drawback in multiple regression. As more variables are included in the analysis, the R^2 value increases (or could remain the same). It is better to use an adjusted R^2, taking into account the number of variables and the sample size. The formula for the adjusted R^2 is given in (7.12).

$$Adjusted \ R^2 = 1 - \frac{(n-1)}{[n-(k+1)]}(1 - R^2) \tag{7.12}$$

Here n represents the sample size, and k represents the number of independent variables.

Adjusted R^2 for the Example 7.5 is (Adjusted R-squared: 0.858) as shown at the bottom of the results of summary() function.

Adjusted $R^2 = 1 - \frac{n-1}{[n-(k+1)]}\left(1 - R^2\right) = 1 - \frac{7-1}{[7-(1+1)]}(1 - 0.88) = 0.858$

Example 7.7: Chemical composition—The chemical composition of black tea infusions and their relationship with the sensory quality assessed by tea tasters were analyzed for a total quality score (Y), the theaflavin-3′-gallate (X_1) and an infusion color indicator (X_2). The data (arbitrary data) are given in Table 7.4.

The data had been stored as a .CSV (Example7_7) file. R provides the function lm() to fit a simple linear regression model and a multiple linear regression model. The results of using R commands to fit the regression model for the chemical composition data in Table 7.4 are given below.

```
> model <- lm(y ~ x1 + x2, data = Example7_7)
> model

Call:
lm(formula = y ~ x1 + x2, data = Example7_7)

Coefficients:
(Intercept)          x1           x2
   68.28574     2.42395      0.03832
```

TABLE 7.4 Arbitrary Data for the Chemical Composition

Y	X_1	X_2
75.30	1.45	−41.33
55.20	1.20	−29.51
77.25	1.89	−27.37
67.51	0.95	−29.17
80.27	0.85	−25.33
67.20	3.21	−29.11
83.20	5.20	−25.73
71.45	2.95	−35.15
83.55	7.58	−35.23
79.77	4.97	−41.67
82.56	5.11	−35.12

The results of using `lm()` are only three values that represent the coefficient of the regression model. The first value represents the intercept ($b_0 = 68.28574$), the second value represents the slope of the first variable X_1 ($b_1 = 2.42395$), and the last result is for the slope of the second independent variable X_2 ($b_2 = 0.03832$). The model that describes the relationship between the total quality score (Y) and the two independent variables for the theaflavin-3′-gallate (X_1) and the infusion color indicator (X_2) can be summarized in a mathematical equation similar to (7.10).

$$\hat{Y} = 68.286 + 2.424X_1 + 0.038X_2$$

We tested the significance of the model by producing the ANOVA Table, which includes the degrees of freedom, sum of squares, mean sum of squares, F-value, and p-value. Moreover, we produced a t-test for the individual variable to make a decision regarding the effect of each independent variable on the response (quality scores).

```
> anova (lm(y ~ x1 + x2, data = Example7_7)) # anova table
Analysis of Variance Table

Response: y
          Df Sum Sq Mean Sq F value  Pr(>F)
x1         1 282.70 282.697  4.6029  0.06423 .
x2         1   0.45   0.447  0.0073  0.93410
Residuals  8 491.34  61.417
---
Signif. codes: 0 '***' 0.001 '**' 0.01 '*' 0.05 '.' 0.1 ' ' 1
> summary (lm(y ~ x1 + x2, data = Example7_7)) # show results

Call:
lm(formula = y ~ x1 + x2, data = Example7_7)

Residuals:
    Min      1Q   Median      3Q     Max
-14.864  -2.300    1.034   4.190  10.895
```

(Continued)

(Continued)

```
Coefficients:
             Estimate  Std. Error  t value  Pr(>|t|)
(Intercept)  68.28574   13.97376    4.887   0.00121 **
x1            2.42395    1.17464    2.064   0.07295 .
x2            0.03832    0.44917    0.085   0.93410
---
Signif. codes: 0 '***' 0.001 '**' 0.01 '*' 0.05 '.' 0.1 ' ' 1

Residual standard error: 7.837 on 8 degrees of freedom
Multiple R-squared: 0.3656, Adjusted R-squared: 0.207
F-statistic: 2.305 on 2 and 8 DF, p-value: 0.162
```

The results of the analysis showed that the model is not significant. This result indicates that none of the selected independent variables (X_1 and X_2) affect the dependent variable (Y), as the p-values ($Pr(>F)$) for X_1 and X_2 are large: the p-value < 0.06423 and the p-value < 0.93410 for X_1 and X_2, respectively. Furthermore, the effect of X_1 can be considered an influential factor on the response at a p-value < 0.08. As the ANOVA results exhibited an insignificant effect from the independent variables, no further analysis should be explored, and there is no need for the t-test.

7.7 MULTIVARIATE MULTIPLE LINEAR REGRESSION

Multiple regression can easily be extended to address situations where there is more than one response variable (dependent, output) and more than one independent variable. Then the model is called a multivariate multiple regression model. Multivariate means that there is more than one response, and multiple means there is more than one independent variable. To model the relationship between p responses and k independent variables, there should be p regression models, one for each response as shown below.

$$\hat{Y}_1 = b_{01} + b_{11}X_1 + \ldots + b_{k1}X_k$$

$$\hat{Y}_2 = b_{02} + b_{12}X_1 + \ldots + b_{k2}X_k$$

$$\vdots$$

$$\hat{Y}_p = b_{0p} + b_{1p}X_1 + \ldots + b_{kp}X_k$$

The matrix formula is similar to the ordinary least squares formula, with the only difference being that Y is a matrix response variable and not a vector as given in (7.13).

$$b = (X'X)^{-1}X'Y \tag{7.13}$$

where,

$$Y = \begin{bmatrix} Y_{11} & Y_{12} & \ldots & Y_{1p} \\ Y_{21} & Y_{22} & \ldots & Y_{2p} \\ \vdots & \vdots & & \vdots \\ Y_{n1} & Y_{n2} & \ldots & Y_{np} \end{bmatrix} = \begin{bmatrix} Y'_1 \\ Y'_2 \\ \vdots \\ Y'_n \end{bmatrix}$$

$$X = \begin{bmatrix} 1 & X_{11} & X_{12} & \ldots & X_{1k} \\ 1 & X_{21} & X_{22} & \ldots & X_{2k} \\ \vdots & \vdots & \vdots & \ldots & \vdots \\ 1 & X_{n1} & X_{n2} & \ldots & X_{nk} \end{bmatrix}$$

$$\boldsymbol{\beta} = \begin{bmatrix} \beta_{01} & \beta_{02} & \cdots & \beta_{0p} \\ \beta_{11} & \beta_{12} & \cdots & \beta_{1p} \\ \vdots & \vdots & \cdots & \vdots \\ \beta_{k1} & \beta_{k2} & \cdots & \beta_{kp} \end{bmatrix}$$

Note
- Each response follows its regression model, and thus the procedure for estimating the coefficient of multiple linear regression models is used for each model in the case of multivariate multiple regression models.

FURTHER READING

Alkarkhi, A. F. M., & Low, H. C. (2012). *Elementary statistics for technologist*. Pulau Pinang, Malaysia: Universiti Sains Malaysia press.

Alkarkhi, A. F. M., Norli., & Azhar, M. E. (2008). Assessment of arsenic and heavy metal contents in cockles (*Anadara granosa*) using multivariate statistical techniques. *Journal of Hazardous Materials, 150,* 783−789.

Chi, Y. (2009). *R Tutorial, An Introduction to statistics - Estimated Simple Regression Equation* [Online]. Available: <http://www.r-tutor.com/elementary-statistics/simple-linear-regression/estimated-simple-regression-equation> Accessed 07.03.16.

Daniel, H. (2013). *Blog Archives, High Resolution Figures in R* [Online]. Available: <https://www.r-bloggers.com/author/daniel-hocking/> Accessed 15.07.17.

Draper, N. R., & Smith, H. (1998). *Applied regression analysis*. New York: Wiley.

Foroogh, B., Alkarkhi, A. F. M., & Azhar, M. E. (2008). Antioxidant activity and phenolic content of various date palm (*Phoenix dactylifera*) fruits from Iran. *Food Chemistry, 107,* 1636−1641.

Johnson, R. A., & Wichern, D. W. (2002). *Applied multivariate statistical analysis*. New Jersey: Prentice Hall.

Kabacoff, R. I. (2014). *Quick R - accessing the power of R - Multiple (Linear) Regression* [Online]. Available: <http://www.statmethods.net/stats/regression.html> Accessed 05.05.16.

Rencher, A. C. (2002). *Methods of multivariate analysis*. New York: J. Wiley.

Chapter 8

Principal Components Analysis

Learning outcomes

At the end of this chapter, you should be able

- *To describe principal components analysis (PCA).*
- *To know how to apply PCA for data analysis.*
- *To apply R commands and built-in functions for PCA.*
- *To understand and interpret the results of R output for PCA.*
- *To write a report on the useful conclusions drawn from PCA.*

8.1 INTRODUCTION

Principal component analysis (PCA) is a multivariate technique used to reduce the number of dimensions (data reduction technique) to explain the total variation in the data with a few linear combinations of the original variables, called components, that are uncorrelated. PCA gives the best results when the variables under study are highly correlated (positively or negatively). For example, if there are 15 highly correlated variables, it is quite possible that only two principal components can describe the variables under study. In PCA, there is only one set of variables, while regression analysis addresses two sets of variables. One set is called the dependent variable, and the second set is known as the independent variable, which consists of one or more variables.

8.2 DESCRIBING PRINCIPAL COMPONENTS

It is a good practice to provide the arrangement of the data for a PCA before describing how to calculate the principal components. The arrangement of the data for PCA is given in Table 8.1.

Consider k random variables (Y) measured from each sample for n samples, as shown in Table 8.1. PCA does not require a multivariate assumption; thus, there is no need to test for a normality assumption. The procedure is to find a few components that describe most of the variation in the data, i.e., using the k random variables $Y_1, Y_2, \ldots, Y_k$ to produce components $Z_1, Z_2, \ldots, Z_k$ that are uncorrelated. The k components can be ordered so that the first principal component Z_1 represents the linear combination with the maximum variance, the second principal component Z_2 represents the second largest amount of variation, which is less than the amount explained by Z_1, and so on for other components. The principal components can be arranged in order based on the amount of variance explained by each component, var $(Z_1) \geq \text{var}(Z_2) \geq \ldots \geq \text{var}(Z_k)$. Researchers are expected to describe the variation in the data with the first few components and ignore other components as they explain a minimal amount of the variation.

The principal components are represented by the linear combination of all variables included in the study $Y_1, Y_2, \ldots, Y_k$ as shown below.

$$Z_1 = a_{11}Y_1 + a_{12}Y_2 + a_{13}Y_3 + \ldots + a_{1k}Y_k$$
$$Z_2 = a_{21}Y_1 + a_{22}Y_2 + a_{23}Y_3 + \ldots + a_{2k}Y_k$$
$$\vdots$$
$$Z_k = a_{k1}Y_1 + a_{k2}Y_2 + a_{k3}Y_3 + \ldots + a_{kk}Y_k$$

where the $a's$ are the coefficients of the principal component.

Easy Statistics for Food Science with R. DOI: https://doi.org/10.1016/B978-0-12-814262-2.00008-X
© 2019 Elsevier Inc. All rights reserved.

TABLE 8.1 The General Arrangement of Data for Principal Components Analysis

Variable sample No.	Y_1	Y_2	$\ldots$	Y_k
1	Y_{11}	Y_{12}	$\ldots$	Y_{1k}
2	Y_{21}	Y_{22}	$\ldots$	Y_{2k}
$\vdots$	$\vdots$	$\vdots$	$\ldots$	$\vdots$
n	Y_{n1}	Y_{2n}	$\ldots$	Y_{n2}

The first principal component is represented by the first linear combination (Z_1) that maximizes var(Z_1) (var(Z_1) is as large as possible), subject to:

$$a_{11}^2 + a_{12}^2 + \ldots + a_{1k}^2 = 1$$

This condition is critical because the variance of Z_1 can be raised by increasing any value of a_{1i} (such as multiplying by a constant) in the linear combination. The second principal component is represented by the second linear combination (Z_2), which maximizes var(Z_2) (var(Z_2) is as large as possible) subject to

$$a_{21}^2 + a_{22}^2 + \ldots + a_{2k}^2 = 1$$

Z_1 is uncorrelated with Z_2, which means cov(Z_1,Z_2) = 0.
Other principal components are defined in a similar manner until k uncorrelated components are obtained.

8.3 CALCULATING PRINCIPAL COMPONENTS

It is better to have a general idea of the procedure for calculating the principal components than to discuss how the principal components equations are derived because the calculation for finding principal components is carried out using a statistical program. The steps for calculating the principal components from covariance matrix are given below.

- The first step in calculating the principal components is to calculate the covariance matrix between different variables. The values allocated on the diagonal of the matrix represent the variance, while off-diagonal values represent the covariance of the variables.
- The second step is to calculate the eigenvalues λ_i of the covariance matrix; the number of eigenvalues is equal to the number of variables and can be arranged in order:

$\lambda_1 \geq \lambda_2 \geq \ldots \geq \lambda_k \geq 0$. The eigenvalues represent the variances of the principal components Z_i. λ_1 represents the variance of the first principal components Z_1, λ_2 represents the variance of the second principal component Z_2, and so on.

- After calculating the eigenvalues, the eigenvectors associated with each eigenvalue are calculated to represent the coefficients of the principal components $a_1, a_2, a_3, \ldots, a_k$. The elements of each eigenvector are called loadings. If the eigenvalues are distinct, the k eigenvectors are orthogonal.
- Finally, select the first few components that account for most of the variation in the data to represent the data.

Note:
- The sum of all the eigenvalues is equal to the sum of the diagonal elements of the covariance matrix, which is equal to the sum of the variances of the original variables.

$$\sum_{i=1}^{k} S_{ii} = \lambda_1 + \lambda_2 + \ldots + \lambda_k$$

Where S_{ii} represents the variance of the variables and is presented on the diagonal of the covariance matrix.
- The loadings a_i measures the importance of the variable to the principal component, regardless of other variables.
- Each component measures different dimension in the dataset.

8.4   PRINCIPAL COMPONENTS FROM CORRELATION MATRIX

In some cases, a correlation matrix is used to extract the components instead of using a covariance matrix, such as unequal variances or when the measurement units are different (not commensurate). In this case, the variables with large variances will contribute more to the component than variables with small variances. Another reason to use a correlation matrix rather than a covariance matrix is that correlation matrix is an invariant scale, which means that changing the scales of the variables will not influence the components. Furthermore, using a correlation matrix will make the components more interpretable. The steps for calculating the principal components from correlation matrix are the same to the covariance matrix.

8.5   STANDARDIZING THE PRINCIPAL COMPONENTS

Some studies measure the variables on different scales due to their nature. To avoid the effect of unequal variances of the variables, it is better to standardize the variables and render them with invariant scales as well. Standardization can be accomplished by using the following formula:

$$Z = \frac{y_i - \overline{y}}{s_i}$$

The new matrix (standardized values) is similar to the correlation matrix.

8.6   CHOOSING THE NUMBER OF PRINCIPAL COMPONENTS

It is important to carefully choose the number of the components that should remain in the analysis as this is an important decision. The goal of using principal components is to summarize the data and reduce its dimensionality; proper selection will reflect these goals. The following are some tips on how to choose the most essential components:

- Keep the components whose eigenvalues are greater than 1 when using a correlation matrix to extract the components. For a covariance matrix, keep the components whose eigenvalues are greater than the average.
- Use the plot of λ_i versus i (called a Scree plot) and find the break between the large and small eigenvalues.

8.7   PRINCIPAL COMPONENTS ANALYSIS IN R

PCA requires some packages to be installed from the R library, such as `GPArotation,psych` and other packages. R provides a variety of functions from different packages to conduct PCA. The following are the necessary functions, and we also give the structure of each function and their corresponding packages.

1. The function `princomp()` is used to calculate eigenvalues, and it is available in the package `stats`. This function comes by default with `stats`.

```
princomp (data frame, cor = TRUE)
```

Whereas `cor = TRUE` means calculate the eigenvalues from the correlation matrix, `cor = FALSE` means calculate the eigenvalues from the variance-covariance matrix. This function will be used to analyze the examples in this chapter. The reader can try the other functions given in this chapter to conduct PCA. The functions used with `princomp` to extract the results are given below.

2. The function `summary()` is used to provide both a proportion of the variance in the data explained by each component and the cumulative proportion of the explained variance.

```
summary (data frame)
```

3. The function `$loadings` is used to produce the loadings for the components.

```
Name of the PCA $ loadings
```

4. The function `$scores` is used to produce the scores (the values of the principal components) for all observations.

```
Name of the PCA $ scores
```

5. The function `screeplot()` is used to produce the Scree plot for the extracted eigenvalues.

```
screeplot (data frame, npcs = number of components, ...)
```

6. The function `$sdev` is also used to find the eigenvalues.

```
Name of PCA $ sdev
```

There are additional functions that can be used for performing PCA, and they are given below:

7. The `prcomp()` function is used to calculate the eigenvalues associated with each component; this function is available in the package `Stats` and comes by default.

```
prcomp (data frame, scale = TRUE )
```

The structure `scale = TRUE` means calculate the PCA from the correlation matrix. Other functions are used to extract the results, e.g., the function `$x` is used to extract the scores, and the function `Name$rotation` is used to extract the loadings.

8. The function `PCA()` is available in the package (`FactoMineR`).

```
PCA (data frame, graph = TRUE)
```

The structure `graph = TRUE` indicates a Scree plot. The function `$eig` is used to extract the results for the eigenvalues' percentage of the cumulative variance and the percentage of the variance, and the function `$svd` is used to extract the scores. Other functions are associated with this call and can be seen from the results of using the function `PCA()`.

9. The function `dudi.pca()` is available in the package `ade4`.

```
dudi.pca (data frame, scale = FALSE, scannf = FALSE, nf = No of components)
```

The structure `scannf = FALSE` and `nf = No of components` are for the Scree plot and to select the number of components. The function `$c1` is used to extract the loadings, and the function `$eig` is used to extract the scores.

10. The function `acp()` is available in the package `amap`.

```
acp (data frame)
```

The results produced by different functions will be the same, including the eigenvalues, proportions of variance, cumulative variances, loadings, scores, and related information. The function $eig is used to extract the eigenvalues, and the function $loadings is used to extract the loadings. The last function is $scores.

Note:

- There are other functions to produce the eigenvalues, loadings, and scores based on the commands used to carry out PCA.

Example 8.1: The physicochemical properties of bananas—A researcher wants to investigate the physicochemical properties of banana pulp and peel flour prepared from green and ripe Cavendish bananas. The researcher measures the physicochemical properties such as the pH, total soluble solids (TSS), and water holding capacity (WHC) and oil holding capacity (OHC) at 40°C, 60°C, and 80°C. In addition, the researcher measures the color values L*, a* and b*, the back extrusion force (texture = BEF), and the viscosity. The four types of flour are green peel (Gpe), ripe peel (Rpe), green pulp (Gpu), and ripe pulp (Rpu). The data are given in Table 1.2.

The variances of different random variables (parameters) should first be checked to obtain a general picture of the dispersion of the data for each variable. The function sapply() in R was used to calculate the variances of the 13 variables included in this study. The results of the variances produced by R were rounded to two decimals using the function round(). The data had been stored as a .CSV (Example8_1) file.

```
> round (sapply (Example8_1[2:14], var), digits = 2)
    pH     TSS       L       a       b    WHC40    WHC60    WHC80
   0.19    1.60  298.62    2.14   21.25     3.16     3.34     1.82
  OHC40    OHC60    OHC80   Viscosity   Texture
   0.01     0.03     0.03      360.45      503.46
```

The results showed that the variances of the texture (503.46), viscosity (360.45), and L* (298.62) are very large compared to the other variances, which means that these three variables would dominate the principal components. Thus, it is better to standardize the variables, which is equivalent to using the correlation matrix to calculate the eigenvalues. The correlation matrix for the physicochemical variables was calculated using the function corr.test(). This function was used rather than cor() since cor()does not test the significance of the correlation.

```
Corr.test(Example8_1[2:14]), digits = 2)
> corr.test(Example8_1[2:14])
Call:corr.test(x = Example8_1[2:14])
Correlation matrix
          pH   TSS    L     a     b WHC40 WHC60 WHC80 OHC40 OHC60 OHC80  Viscosity
Texture
pH      1.00  0.34 -0.05 -0.12 -0.41  0.14  0.10  0.36  0.22  0.30  0.30      0.31
0.18
TSS     0.34  1.00 -0.03  0.11 -0.64 -0.38 -0.59 -0.11  0.25  0.72  0.58      0.97
0.08
L      -0.05 -0.03  1.00 -0.90 -0.13 -0.77 -0.53 -0.57 -0.34 -0.66 -0.65     -0.11
-0.95
a      -0.12  0.11 -0.90  1.00  0.08  0.66  0.38  0.50  0.32  0.69  0.69      0.20
0.85
b      -0.41 -0.64 -0.13  0.08  1.00  0.11  0.16 -0.25 -0.60 -0.39 -0.42     -0.54
0.04
WHC40   0.14 -0.38 -0.77  0.66  0.11  1.00  0.92  0.86  0.41  0.26  0.41     -0.34
0.81
WHC60   0.10 -0.59 -0.53  0.38  0.16  0.92  1.00  0.82  0.34 -0.05  0.13     -0.57
0.58
```

(Continued)

(Continued)

```
WHC80      0.36 -0.11 -0.57 0.50 -0.25  0.86  0.82  1.00  0.56 0.34  0.50      -0.09
0.65
OHC40      0.22  0.25 -0.34 0.32 -0.60  0.41  0.34  0.56  1.00 0.41  0.60       0.21
0.37
OHC60      0.30  0.72 -0.66 0.69 -0.39  0.26 -0.05  0.34  0.41 1.00  0.86       0.74
0.67
OHC80      0.30  0.58 -0.65 0.69 -0.42  0.41  0.13  0.50  0.60 0.86  1.00       0.61
0.70
Viscosity  0.31  0.97 -0.11 0.20 -0.54 -0.34 -0.57 -0.09  0.21 0.74  0.61       1.00
0.13
Texture    0.18  0.08 -0.95 0.85  0.04  0.81  0.58  0.65  0.37 0.67  0.70       0.13
1.00
Sample Size
[1] 48
```

Probability values (Entries above the diagonal are adjusted for multiple tests.)

	pH	TSS	L	a	b	WHC40	WHC60	WHC80	OHC40	OHC60	OHC80	Viscosity	Texture
pH	0.00	0.58	1.00	1.00	0.17	1.00	1.00	0.46	1.00	1.00	1.00	0.86	1.00
TSS	0.02	0.00	1.00	1.00	0.00	0.31	0.00	1.00	1.00	0.00	0.00	0.00	1.00
L	0.74	0.85	0.00	0.00	1.00	0.00	0.01	0.00	0.57	0.00	0.00	1.00	0.00
a	0.42	0.46	0.00	0.00	1.00	0.00	0.31	0.01	0.76	0.00	0.00	1.00	0.00
b	0.00	0.00	0.39	0.57	0.00	1.00	1.00	1.00	0.00	0.24	0.12	0.00	1.00
WHC40	0.33	0.01	0.00	0.00	0.46	0.00	0.00	0.00	0.17	1.00	0.15	0.61	0.00
WHC60	0.49	0.00	0.00	0.01	0.29	0.00	0.00	0.00	0.61	1.00	1.00	0.00	0.00
WHC80	0.01	0.47	0.00	0.00	0.08	0.00	0.00	0.00	0.00	0.57	0.02	1.00	0.00
OHC40	0.14	0.09	0.02	0.03	0.00	0.00	0.02	0.00	0.00	0.15	0.00	1.00	0.35
OHC60	0.04	0.00	0.00	0.00	0.01	0.07	0.76	0.02	0.00	0.00	0.00	0.00	0.00
OHC80	0.04	0.00	0.00	0.00	0.00	0.00	0.37	0.00	0.00	0.00	0.00	0.00	0.00
Viscosity	0.03	0.00	0.47	0.18	0.00	0.02	0.00	0.56	0.15	0.00	0.00	0.00	1.00
Texture	0.23	0.61	0.00	0.00	0.78	0.00	0.00	0.00	0.01	0.00	0.00	0.38	0.00

To see confidence intervals of the correlations, print with the short = FALSE option

The correlation matrix for the physicochemical properties of the bananas was examined. It was found that some variables exhibited a strong correlation (positive or negative) with other variables. For instance, whereas TSS showed a strong positive relationship with pH, OHC60, OHC80, and viscosity, a strong negative relationship was exhibited with b*, WHC40, and WHC60. This finding indicates that these variables have a strong association with the other variables and share an original source.

PCA was carried out on the dataset (with 13 variables) to identify the source of the variation between the pulp and peel and between the stage of ripeness. The `princomp (data frame, cor = TRUE)` and `summary()` functions were used to perform the PCA. The command `cor = TRUE` means calculate the principal components from the correlation matrix, and `cor = FALSE` would have R calculate the principal components from the covariance matrix.

```
pca <- princomp (Example8_1[2:14], cor = TRUE)
summary (pca)
```

The function `princomp()` was used to calculate the standard deviation (the square root of the eigenvalue) associated with each principal component. Furthermore, the function `summary(pca)` was used to calculate the proportion of the variance in the data explained by each component and the cumulative proportion of the explained variance. The results of the R commands we employed to investigate the physicochemical properties of banana data extracted 13 components, including the square root of the eigenvalue, which is represented by the `standard deviation`, the `proportion of the variance` explained by each component and the `cumulative proportion`.

```
> pca < - princomp (Example8_1[2:14], cor = TRUE)
> summary (pca)
Importance of components:
                              Comp.1       Comp.2       Comp.3        Comp.4
Standard deviation         2.4184511    1.9641988    1.2976567     0.9129025
Proportion of Variance     0.4499158    0.2967751    0.1295318     0.0641070
Cumulative Proportion      0.4499158    0.7466910    0.8762227     0.9403297
                              Comp.5       Comp.6       Comp.7        Comp.8
Standard deviation         0.5173722    0.3999367    0.338429569   0.31441276
Proportion of Variance     0.020590     0.0123038    0.008810352   0.00760426
Cumulative Proportion      0.9609201    0.9732238    0.982034200   0.98963846
                              Comp.9      Comp.10       Comp.11       Comp.12
Standard deviation         0.24119430   0.1828193    0.146787968   0.10622599
Proportion of Variance     0.004474976  0.0025709    0.0016574     0.000867
Cumulative Proportion      0.99411344   0.99668443   0.998342      0.9992099
                             Comp.13
Standard deviation         0.1013496950
Proportion of Variance     0.0007901354
Cumulative Proportion      1.0000000000
```

Observe that the first three components explain more than 87% of the total variance in the data. The first component explains 44.99158% of the total variance $[((2.4184511)^2/13) = (5.849/13) = 44.99158\%]$, which is the highest contribution, and the second principal component explains 29.67751% $[((1.9641988)^2/13) = (3.858/13) = 29.67751\%]$ of the total variance, which is the second-highest contribution. The other components can be explained in a similar manner.

The next step is to decide how many components are necessary to explain the variation in the data. Applying the steps presented in Section 8.6 on the summary results will give three components with eigenvalues greater than 1. (An eigenvalue gives a measure of the significance of the component; the component with the highest eigenvalue is the most significant and is responsible for explaining the largest amount of variation in the data.) The Scree plot for the eigenvalues can be produced by using the function `screeplot()` in R to obtain a visual picture of the different eigenvalues. The Scree plot for the physicochemical properties of bananas using R commands is given in Fig. 8.1.

```
screeplot(pca, npcs = 13, type = "barplot", xaxt = "n", yaxt = "n")
```

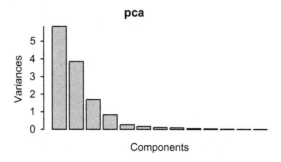

FIGURE 8.1   Scree plot for the extracted eigenvalues calculated from the physicochemical properties of banana data.

This command will produce the Scree plot as a bar chart. It is possible to produce a Scree plot with connected lines by changing the type to lines with `type = "lines"`. In addition, `npcs` is used to identify the number of components to be plotted.

It is clear from Fig. 8.1 that three components are enough to capture most of the variance in the data.

The function `$loadings` is usually used to calculate the loadings for each component.

```
Name of the PCA $ loadings
```

However, excessively small coefficients will be missing in the results of the loadings even if they are not necessarily equal to zero. The function `unclass()` can be used to show all of the coefficients. The name of the place where the PCs are stored is `pca`.

```
Unclass (pca $ loadings)
round (unclass (pca $ loadings), digits = 3)
```

The loadings for all components were calculated using the function `pca$loadings` in R, and the function `unclass` was used to display all of the values. The results of applying R commands to calculate the loadings yielded 13 components. The R outputs were rounded to three decimals.

```
> round (unclass (pca $ loadings), digits = 3)
```

	Comp.1	Comp.2	Comp.3	Comp.4	Comp.5	Comp.6	Comp.7
pH	0.125	-0.153	0.432	0.746	0.333	0.056	-0.090
TSS	0.109	-0.484	-0.045	0.042	-0.136	0.006	0.167
L	-0.362	-0.118	0.273	-0.060	-0.164	-0.380	-0.168
a	0.343	0.065	-0.360	-0.135	-0.107	-0.016	-0.145
b	-0.112	0.327	-0.456	0.281	0.389	-0.521	0.352
WHC40	0.320	0.310	0.093	0.052	-0.081	0.032	-0.100
WHC60	0.221	0.390	0.245	0.021	-0.143	-0.038	0.103
WHC80	0.322	0.166	0.333	0.035	-0.472	-0.374	0.322
OHC40	0.258	-0.078	0.367	-0.544	0.621	-0.035	0.294
OHC60	0.325	-0.264	-0.190	0.075	-0.060	0.151	0.135
OHC80	0.354	-0.199	-0.054	-0.075	0.143	-0.539	-0.635
Viscosity	0.125	-0.464	-0.120	0.084	-0.125	-0.230	0.396
Texture	0.378	0.104	-0.178	0.151	0.075	0.265	-0.047

	Comp.8	Comp.9	Comp.10	Comp.11	Comp.12	Comp.13
pH	0.296	-0.040	0.017	-0.040	-0.060	-0.033
TSS	-0.202	-0.024	0.247	0.080	0.114	-0.763
L	0.008	0.156	0.662	-0.292	-0.019	0.168
a	0.717	-0.173	0.232	-0.216	-0.175	-0.149
b	-0.047	0.122	0.115	0.032	0.047	-0.122
WHC40	-0.041	-0.100	0.189	-0.044	0.849	0.037
WHC60	-0.253	0.089	-0.191	-0.679	-0.235	-0.286
WHC80	0.117	0.046	0.001	0.487	-0.181	0.067
OHC40	0.077	0.004	0.132	0.003	-0.006	0.009
OHC60	0.029	0.812	0.017	-0.123	0.071	0.241
OHC80	-0.221	0.033	-0.241	0.057	-0.040	-0.013
Viscosity	-0.057	-0.443	-0.168	-0.361	0.120	0.387
Texture	-0.466	-0.229	0.506	0.105	-0.346	0.239

The objective of PCA is to reduce the dimension of the data by expressing the variables in the form of a linear combination. In this case, it appears that three components can explain more than 87% of the total variance (their eigenvalues are greater than 1). The other components contribute very little to explaining the variation in the data, as shown in the summary() results.

The first three principal components used X_i in the same sequence presented with RStudio results. The three principal components in equation form are given in Eqs. (8.1−8.3). The resulting loadings have the same form as the principal components instead of the original names of the variables; for instance, X_1 represents pH, X_2 represents TSS, and so on for the other parameters.

$$Z_1 = 0.125X_1 + 0.109X_2 - 0.362X_3 + 0.343X_4 - 0.112X_5 + 0.320X_6 + 0.221X_7 + 0.322X_8$$
$$+ 0.258X_9 + 0.325X_{10} + 0.354X_{11} + 0.125X_{12} + 0.378X_{13} \quad (8.1)$$

$$Z_2 = -0.153X_1 - 0.484X_2 - 0.118X_3 + 0.065X_4 + 0.327X_5 + 0.310X_6 + 0.390X_7$$
$$+ 0.166X_8 - 0.078X_9 - 0.264X_{10} - 0.199X_{11} - 0.464X_{12} + 0.104X_{13} \quad (8.2)$$

$$Z_3 = 0.432X_1 - 0.045X_2 + 0.273X_3 - 0.360X_4 - 0.456X_5 + 0.093X_6 + 0.245X_7 + 0.333X_8$$
$$+ 0.367X_9 - 0.190X_{10} - 0.054X_{11} - 0.120X_{12} - 0.178X_{13} \quad (8.3)$$

The contribution of each variable to the first principal component can be measured by the coefficients (loadings) associated with each variable in the first principal component. The contribution of each variable can be arranged in order: BEF (X_{13}) > L* (X_3) > OHC80 (X_{11}) > a* (X_4) > OHC60 (X_{10}) > WHC80 (X_8) > WHC40 (X_6) > WHC40 (X_9) > OHC60 (X_7) > pH (X_1), viscosity (X_{12}) > b* (X_5) > TSS (X_2). The contributions of pH, viscosity, b*, and TSS do not substantially affect the first component relative to the contributions of the other variables. Furthermore, whereas a positive sign means that the variable has the ability to increase the value of the principal component, a negative sign indicates this variable's ability to decrease the value of the first component. The first component represents the difference between the effects of L* (X_3) and b* (X_5) and the combined effects of all the other variables. This difference explains 44.99% of the total variance. The other components can be explained in a similar manner.

It is important to calculate the scores (the values of the principal component for each of the n observations) of the principal component to be used for other purposes, such as to draw a bar chart or scatter diagram for further analysis. The values of the principal components are calculated based on standardized values. For instance, the value of the principal component for the first individual is 0.280. First, the variables should be standardized by using the following formula:

$$Z = \frac{y_i - \bar{y}}{s_i}$$

For instance, the mean and standard deviation for the pH are 5.1117 and 0.4413, respectively. The standardized value for the pH is

$$pH = X_1 = \frac{4.64 - 5.1117}{0.4413} = -1.06874$$

The first principal component of the first sample (individual) is $0.2769 = 0.280$.

$$Z_1 = 0.125(-1.06874) + 0.109(-0.66224) - 0.362(-0.62058) + 0.343(0.788432) - 0.112(0.910569)$$
$$+ 0.320(0.602826) + 0.221(0.261095) + 0.322(-0.41676) + 0.258(-0.42981) + 0.325(-0.47401)$$
$$+ 0.354(0.287763) + 0.125(-0.25337) + 0.378(0.443835) = 0.27675 = 0.280$$

The values of the principal components for the other individuals can be calculated similarly. The scores for all of the observations can easily be calculated using R commands. The function $scores was used to calculate the scores for the physicochemical properties of bananas.

```
round (pca $ scores, digits = 3)
```

The call to pca$scores gives an order to extract ($) the scores from the file pca. The scores for all of the observations are rounded to three decimals as shown below.

```
> round (pca $ scores, digits = 3) # the principal components
```

	Comp.1	Comp.2	Comp.3	Comp.4	Comp.5	Comp.6	Comp.7	Comp.8
[1,]	0.280	1.406	-1.466	-0.398	0.084	-0.187	-0.370	-0.108
[2,]	0.394	1.957	-2.188	-0.877	-0.055	-0.106	-0.235	0.189
[3,]	0.736	1.543	-2.098	-0.504	0.346	-0.554	-0.581	0.339
[4,]	0.525	1.905	-1.828	-0.327	-0.157	0.298	-0.193	0.312
[5,]	0.425	2.109	-2.096	0.019	-0.340	-0.012	0.564	0.172
[6,]	0.450	1.922	-2.052	-0.766	0.088	-0.095	0.394	-0.264
[7,]	0.264	1.913	-2.512	-0.254	-0.730	-0.059	-0.066	-0.286
[8,]	0.602	1.899	-1.190	0.727	0.257	0.531	-0.157	-0.019
[9,]	-0.497	1.581	-1.292	1.534	1.007	-0.264	-0.028	-0.574
[10,]	0.473	1.272	-0.664	0.638	1.349	0.282	0.208	0.063
[11,]	-0.183	1.678	-1.365	1.304	0.890	-0.167	0.548	-0.178
[12,]	0.343	1.357	-1.376	1.236	0.291	0.785	-0.095	0.132
[13,]	3.149	-0.260	-0.027	-1.224	0.109	0.661	-0.519	-0.763
[14,]	3.872	-0.356	1.278	-0.982	0.812	-0.275	0.316	-0.350
[15,]	3.616	-0.818	0.895	-0.575	-0.095	-0.021	0.285	0.075
[16,]	3.546	-0.169	0.918	-0.434	-0.240	0.900	0.908	-0.020
[17,]	3.599	-0.333	0.792	0.166	-0.577	-0.127	0.212	0.148
[18,]	4.407	-0.659	1.596	-0.084	0.189	-0.996	0.253	0.203
[19,]	3.782	-0.211	1.094	0.460	-0.739	-0.333	0.133	0.303
[20,]	4.086	-0.678	1.115	-0.120	0.386	-0.256	0.170	0.382
[21,]	3.573	-0.451	0.245	1.302	-1.171	0.374	-0.102	-0.080
[22,]	4.077	-0.770	0.875	0.291	0.470	-0.193	-0.331	-0.351
[23,]	3.791	-0.274	0.616	0.418	-0.069	0.245	-0.445	-0.141
[24,]	3.735	-0.448	-0.334	0.794	-0.704	0.024	-0.629	0.069
[25,]	-1.948	1.816	1.120	-1.979	0.179	0.217	-0.002	0.078
[26,]	-1.969	1.723	0.926	-2.194	-0.013	0.363	0.184	0.168
[27,]	-1.726	1.614	1.067	-1.330	-0.023	0.333	-0.088	0.484
[28,]	-1.525	1.431	1.733	0.563	0.223	0.699	-0.348	1.024
[29,]	-2.050	1.660	1.365	-1.199	-0.175	-0.349	-0.534	-0.150
[30,]	-2.642	2.116	0.961	-0.804	-0.747	-0.308	-0.093	-0.321
[31,]	-2.452	1.915	0.895	-1.018	-0.636	-0.309	-0.118	-0.364
[32,]	-2.434	1.515	2.440	-0.179	0.796	-0.154	0.000	-0.197
[33,]	-3.034	1.959	1.605	1.482	-0.229	0.140	0.324	0.134
[34,]	-2.815	1.675	1.807	1.462	-0.040	-0.142	0.149	-0.037
[35,]	-2.749	1.712	1.529	1.411	-0.184	-0.476	-0.022	-0.223
[36,]	-2.604	1.744	0.760	1.890	-0.673	-0.293	0.164	-0.144
[37,]	-1.482	-3.453	0.650	0.486	0.341	-0.051	-0.595	-0.058
[38,]	-1.755	-3.269	0.329	0.408	0.609	0.567	-0.282	0.217
[39,]	-1.299	-2.211	-1.911	-0.366	-0.005	-1.088	0.130	0.583
[40,]	-1.210	-3.247	0.299	0.552	0.144	-0.098	-0.539	0.204
[41,]	-1.842	-2.578	-1.200	-0.317	0.327	0.103	0.257	0.387
[42,]	-2.113	-2.641	-1.386	0.208	-0.950	0.091	0.420	0.137
[43,]	-1.723	-3.230	-0.135	-0.181	-0.278	0.127	0.074	-0.242
[44,]	-1.877	-3.159	0.046	-0.844	0.230	0.093	0.547	-0.240
[45,]	-1.358	-2.783	-0.860	-0.148	0.062	-0.372	-0.043	0.181
[46,]	-2.776	-3.020	-0.794	0.144	-0.616	0.470	-0.011	-0.455
[47,]	-1.700	-3.132	-0.364	-0.237	-0.038	-0.094	0.083	-0.150
[48,]	-1.961	-3.272	0.180	-0.157	0.294	0.077	0.104	-0.269

(Continued)

(Continued)

	Comp.9	Comp.10	Comp.11	Comp.12	Comp.13
[1,]	-0.491	-0.109	-0.167	0.207	0.014
[2,]	0.176	0.204	0.204	0.126	-0.114
[3,]	-0.137	0.084	-0.237	0.017	-0.111
[4,]	-0.113	-0.030	0.262	0.016	0.035
[5,]	-0.056	-0.007	0.054	0.081	0.195
[6,]	-0.009	-0.032	-0.072	0.055	0.099
[7,]	-0.107	-0.360	0.109	-0.194	0.137
[8,]	0.266	-0.191	0.021	-0.017	-0.137
[9,]	0.027	0.045	-0.252	-0.109	-0.070
[10,]	-0.080	-0.064	0.240	0.148	0.063
[11,]	0.334	0.222	0.080	-0.290	-0.028
[12,]	-0.134	-0.151	0.169	0.106	0.029
[13,]	0.009	-0.262	-0.194	-0.190	-0.175
[14,]	0.268	-0.266	0.065	0.147	0.038
[15,]	0.430	-0.460	-0.035	0.045	-0.128
[16,]	-0.498	0.186	-0.131	0.010	-0.051
[17,]	0.157	-0.146	-0.202	0.198	0.163
[18,]	-0.095	-0.179	0.037	-0.037	0.010
[19,]	-0.220	-0.069	0.287	-0.084	-0.189
[20,]	-0.133	0.174	-0.098	0.017	-0.087
[21,]	-0.218	0.014	-0.170	-0.230	0.183
[22,]	-0.142	0.377	0.063	-0.079	0.219
[23,]	0.306	0.388	0.041	0.115	-0.111
[24,]	0.214	0.462	0.074	0.046	0.000
[25,]	0.232	0.068	0.024	-0.072	0.022
[26,]	0.292	0.181	-0.080	-0.092	0.046
[27,]	-0.039	0.029	-0.032	0.005	-0.003
[28,]	0.124	-0.212	-0.076	-0.160	0.033
[29,]	-0.433	-0.136	0.060	-0.034	-0.003
[30,]	0.019	0.146	0.029	0.051	0.042
[31,]	0.304	0.139	-0.030	0.010	0.046
[32,]	-0.228	0.131	0.080	0.001	0.009
[33,]	-0.288	0.069	0.085	0.037	-0.102
[34,]	-0.186	-0.019	-0.022	-0.064	-0.096
[35,]	-0.219	-0.109	-0.027	0.100	0.008
[36,]	0.686	-0.102	-0.228	0.122	0.008
[37,]	0.108	-0.060	0.201	0.006	0.047
[38,]	0.082	0.017	-0.247	0.066	0.161
[39,]	-0.155	0.039	-0.134	-0.135	0.003
[40,]	-0.014	-0.212	0.111	-0.017	0.140
[41,]	-0.006	0.146	-0.332	0.033	-0.027
[42,]	0.243	0.134	0.123	-0.005	-0.077
[43,]	0.200	-0.081	0.227	-0.021	-0.002
[44,]	-0.151	0.097	0.136	-0.063	0.022
[45,]	0.035	-0.070	0.015	-0.017	-0.195
[46,]	-0.407	-0.015	-0.106	0.142	-0.171
[47,]	0.037	-0.102	0.010	-0.023	0.067
[48,]	0.010	0.095	0.068	0.028	0.043

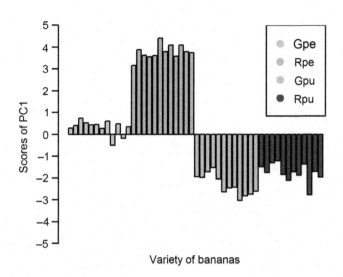

FIGURE 8.2   The values (scores) of the first principal component for the physicochemical properties of bananas.

The scores of the principal components are usually used to construct a bar chart to demonstrate the behavior of different individuals (observations) pictorially. The bar chart for the scores of the first principal component can be produced using built-in functions of R. The bar chart for the scores of the first principal component for the physicochemical properties of bananas is given in Fig. 8.2 (see the appendix for the R code).

```
barplot (A[,1], xaxt = "n", yaxt = "n", ylim = c(-5, 5),
col = c(rep ('gray',12), rep ('darkkhaki',12),rep ('orange',12), rep ('blue',12)))
```

The behavior of different banana samples (individuals) in relation to the first principal component can be studied in Fig. 8.2. Observe that distinctive behavior was exhibited by the samples and based on the banana type, which indicates that the contribution of each sample depends on the variety of banana. For instance, the first 12 samples belong to green peels (Gpe), which showed a positive contribution except for two samples (the positive contribution is mainly due to high values of a*, WHC40, WHC60, WHC80, OCH40, OCH60, OCH80, and the texture, whereas the negative contribution is mainly due to L* and b*. The second 12 bars belong to the ripe peel (Rpe). All samples exhibited a positive contribution and had a larger effect compared to the first variety of Gpe. However, in the next 24 samples, which belong to green pulp (Gpu) and ripe pulp (Rpu), respectively, all of the samples exhibited a negative contribution to the first principal component.

In summary, it can be said that the peel exhibited a positive contribution, which is mainly due to a*, WHC40, WHC60, WHC80, OCH40, OCH60, and OCH80. In contrast, the pulp exhibited a negative contribution to the first principal component, which is mainly due to the high values of L* and b*. Furthermore, the green and ripe samples for both the peel and the pulp exhibited different behavior based on the selected variables. We can also conclude that the stage of ripeness, i.e., whether the banana is green or ripe, affects the behavior of the selected parameters. The bar charts for the scores of the second and third principal components are given in Figs. 8.3 and 8.4, respectively. The interpretation of the other components can be undertaken in a similar manner after accounting for the dominant parameters (high loadings) in each component.

It is useful to depict the scores of the first two components to understand the behavior of the different variables regarding the first two principal components. Fig. 8.5 shows a plot of the values for all of the banana samples (48 samples) for the first two principal components (see the appendix for the R code).

It is clear that there are four different varieties, which indicates that the behavior of each variety is different from the others. Furthermore, the pulp and peel are quite different, and the stage of ripeness has an effect on both the pulp and the peel.

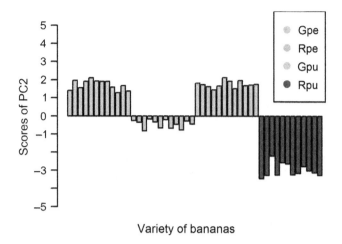

FIGURE 8.3 The values (scores) of the second principal component for the physicochemical properties of bananas.

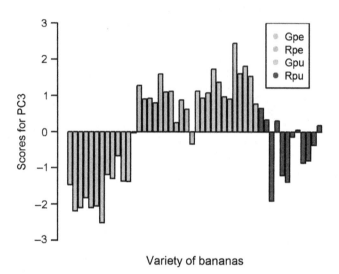

FIGURE 8.4 The values (scores) of the third principal component for the physicochemical properties of bananas.

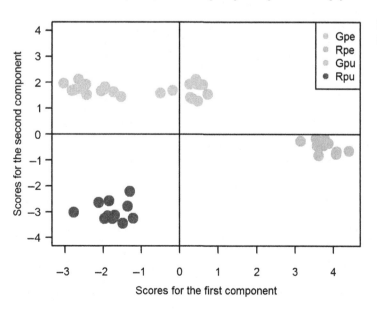

FIGURE 8.5 Plot of the scores of the first and second principal components.

In this example, we have observed how useful principal components are as a data reduction technique and extracted hidden information regarding the behavior of the different variables that are included in the study. Indeed we have summarized 13 variables in only three components that explain more than 87% of the total variation in the data.

Example 8.2: Antioxidants in dates—The edible parts of date palm (*Phoenix dactylifera*) fruits (DPF) were sampled and analyzed for their antioxidant activities (AA) using the Trolox equivalent antioxidant capacity (TEAC) method, 2,2'-azinobis (3-ethylbenzothiazoline-6-sulfonic acid) radical cation (ABTS$^{·+}$) assays and the ferric reducing/antioxidant power method (FRAP assay). The total flavonoid content (TFC) and total phenolic content (TPC) of the DPF were measured using the aluminum chloride colorimetric and Folin—Ciocalteau methods, respectively. We examined four types of soft dates (SD), namely, Jiroft dates, Honey dates, Kabkab dates, and Bam dates; three types of semidry dates (SDD), namely, Piarom dates, Zahedi dates, and Sahroon dates; and one type of dry date (DD) (Kharak dates). The data are given in Table 1.5.

The variances for the antioxidants' activity and the content of the antioxidative compounds in the dates were calculated using the function `sapply()` in R. The data had been stored as a `.CSV` (`Example8_2`) file.

```
round (sapply (Example8_2[2:5], var), digits = 2)
       TFC        TPC AA(Frap)      TEAC
    530.60  1557.33 11353.03   23045.70
```

Notice that the variance of the TFC (`530.60`) is very far from the variance of the other parameters; hence, its contribution to explaining the variation in the data will be very low compared to the contributions of the other parameters. Thus it is better to use a correlation matrix to carry out PCA, which is equivalent to standardizing the values. The correlation matrix for the analysis of the antioxidants in the dates' variables was calculated using the function `corr.test()` in R. The resulting correlation coefficients for the different variables were rounded to two decimals.

```
>corr.test(Example8_2[2:5])
Call:corr.test(x = Example8_2[2:5])
Correlation matrix

          TFC  TPC  AA(Frap)  TEAC
TFC      1.00 0.99      0.99  0.99
TPC      0.99 1.00      1.00  0.97
AA(Frap) 0.99 1.00      1.00  0.97
TEAC     0.99 0.97      0.97  1.00
Sample Size
[1] 24
Probability values (Entries above the diagonal are adjusted for multiple tests.)

          TFC  TPC  AA(Frap)  TEAC
TFC        0    0         0     0
TPC        0    0         0     0
AA(Frap)   0    0         0     0
TEAC       0    0         0     0
To see confidence intervals of the correlations, print with the short = FALSE option
```

The correlation matrix of the analysis of the antioxidants in dates was examined. A strong positive correlation was exhibited between all pairs of variables, which indicates that the correlated parameters have strong associations with each other and share a common source.

PCA was conducted on the dataset (4 variables). Four components were extracted from the correlation matrix. The `summary()` function provided information on the standard deviations, which represent the square roots of the eigenvalues, the proportion of variance explained, and the cumulative proportion for each component.

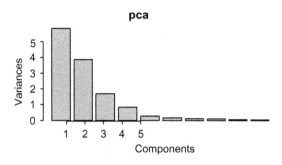

FIGURE 8.6 The Scree plot for the extracted eigenvalues calculated from the antioxidants in dates.

```
> pca <- princomp (Example8_2[2:5], cor = TRUE)
> summary(pca) # print variance accounted for

Importance of components:
                        Comp.1      Comp.2      Comp.3      Comp.4
Standard deviation     1.988440  0.191432352  0.096130259  1.484897e-02
Proportion of Variance 0.988473  0.009161586  0.002310257  5.512295e-05
Cumulative Proportion  0.988473  0.997634620  0.999944877  1.000000e+00
```

PCA yielded only one component with an eigenvalue greater than 1, and it explained more than 98% of the total variance in the data, as shown in the summary results. A Scree plot for the extracted eigenvalues was produced using the function screeplot() in R, and it depicts the difference between the first eigenvalue compared to the others (Fig. 8.6).

The loadings associated with each principal component were calculated using the $loadings and Unclass() functions. The function Unclass()was used to show all of the results from the loadings, as the function princomp() will not display very small values for the loadings.

```
> round (unclass (pca $ loadings), digits = 2) #loadings
          Comp.1  Comp.2  Comp.3  Comp.4
TFC        -0.5    0.20    0.83    0.16
TPC        -0.5   -0.48   -0.05   -0.72
AA(Frap)   -0.5   -0.45   -0.33    0.66
TEAC       -0.5    0.73   -0.45   -0.11
```

The only principal component with an eigenvalue greater than 1 is presented in equation form.

$$Z_1 = -0.5TFC - 0.5TPC - 0.5FRAP - 0.5TEAC \tag{8.4}$$

Thus, only one principal component (8.4) accounted for more than 98% of the total variance, and it was negatively correlated with TFC, TPC, FRAP, and TEAC. This component may be labeled the average of all the selected parameters since all of these parameters contributed highly to explaining the total variation in the data.

The scores of the first principal component for all of the observations were calculated using the function $ scores() in R.

```
> round (pca $ scores, digits = 3)
           Comp.1    Comp.2    Comp.3    Comp.4
  [1,]      0.782    -0.011     0.006    -0.002
  [2,]      0.785    -0.020     0.014     0.000
  [3,]      0.789    -0.003     0.020    -0.010
  [4,]      0.682    -0.019    -0.090     0.004
  [5,]      0.695    -0.010    -0.080    -0.013
  [6,]      0.692    -0.022    -0.046    -0.001
  [7,]      0.804    -0.019     0.020     0.005
  [8,]      0.806    -0.014     0.029    -0.005
  [9,]      0.808    -0.017     0.039     0.005
 [10,]      0.826     0.023    -0.015    -0.013
 [11,]      0.874    -0.038     0.029    -0.009
 [12,]      0.864    -0.029     0.029    -0.006
 [13,]      0.608     0.034    -0.041     0.027
 [14,]      0.637    -0.002    -0.002     0.034
 [15,]      0.625     0.029     0.018     0.025
 [16,]      0.779     0.062    -0.042    -0.026
 [17,]      0.818     0.012    -0.006    -0.027
 [18,]      0.823     0.010     0.009    -0.018
 [19,]      0.648     0.093     0.003     0.006
 [20,]      0.651     0.067     0.070     0.012
 [21,]      0.718     0.003     0.053     0.014
 [22,]     -4.710    -0.433    -0.307     0.003
 [23,]     -5.134    -0.402     0.306    -0.004
 [24,]     -5.870     0.711    -0.016    -0.002
```

The behavior of the antioxidant activity and the content of the antioxidative components for each type of date regarding the principal component can be visualized pictorially using a bar chart. This bar chart permits one to study the relationship between the component scores and the samples from different types of dates. The bar chart for the scores of the principal component for different types of dates is given in Fig. 8.7. Whereas the SDs and SDDs contributed positively to the component, the DDs contributed negatively to the component. This difference in the contribution level indicates that the DDs had different characteristics from the SDs and SDDs.

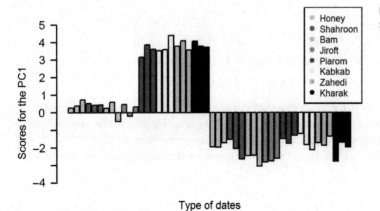

FIGURE 8.7 The scores of the principal components for different types of dates.

FURTHER READING

Abbas, F. M. A., Foroogh, B., Liong, M. T., & Azhar, M. E. (2008). Multivariate statistical analysis of antioxidants in dates (*Phoenix dactylifera*). *International Food Research Journal, 15*, 193–200.

Alkarkhi, A. F. M., Saifullah, B. R., Yeoh, S. Y., & Azhar, M. E. (2011). Comparing physicochemical properties of banana pulp and peel flours prepared from green and ripe fruits. *Food Chemistry, 129*, 312–318.

Bryan, F. J. M. (1991). *Multivariate statistical methods : A primer.* Great Britain: Chapman & Hall.

Daniel, H. (2013). *Blog Archives, High Resolution Figures in R* [Online]. Available: <https://www.r-bloggers.com/author/daniel-hocking/> Accessed 15.07.17.

Gaston, S. (2012). *5 functions to do Principal Components Analysis in R* [Online]. Available: <http://gastonsanchez.com/visually-enforced/how-to/2012/06/17/PCA-in-R/> Accessed 15.07.17.

Ian, H. *dudi.pca: Principal Component Analysis* [Online]. Available: <https://rdrr.io/cran/ade4/man/dudi.pca.html> Accessed 13.08.17.

Johnson, R. A., & Wichern, D. W. (2002). *Applied multivariate statistical analysis.* New Jersey: Prentice Hall.

Kabacoff, R. I. (2014). *Quick R accessing the power of R, Principal Components and Factor Analysis* [Online]. Available: <http://www.statmethods.net/advstats/factor.html> Accessed 07.05.16.

Rencher, A. C. (2002). *Methods of multivariate analysis.* New York: J. Wiley.

Sarah, S. (2012). *Instant R, Performing a principal component analysis in R [Online]. WordPress | Nest Theme by YChong.* Available: <http://www.instantr.com/2012/12/18/performing-a-principal-component-analysis-in-r/> Accessed 11.07.16.

Chapter 9

Factor Analysis

Learning outcomes

At the end of this chapter, you should be able

- *To describe factor analysis (FA).*
- *To know how to apply FA for data analysis.*
- *To apply R commands and built-in functions for FA.*
- *To understand the R output from FA.*
- *To interpret the results and write a report on the useful conclusions.*

9.1 INTRODUCTION

Factor analysis (FA) is a multivariate method and can be considered as an extension of the principal component analysis. The main goal of FA is to describe the relationships among a set of k observable variables with a smaller number of unobservable variables called factors. FA works very well when the variables under study are highly correlated and the goal is to group the variables based on the extracted factors. For instance, if all the variables in one group are highly correlated among themselves and have little correlation with the variables in the remaining groups, each group can represent a factor.

9.2 FACTOR ANALYSIS MODEL

The arrangement of the data for FA is similar to the arrangement of data in principal component analysis (Table 8.1). Suppose a random sample with k observable variables has a mean μ and a covariance matrix $\sum$. Each observable variable of the k variables can be expressed as a linear combination of m ($m < k$) unobservable variables $f_1, f_2, \ldots, f_m$ called common factors and an additional source of variation $\varepsilon_1, \varepsilon_2, \ldots, \varepsilon_k$ called the specific factors (error). The FA model for k observable variables and m common factors is given below.

$$Y_1 = d_{11}f_1 + d_{12}f_2 + \ldots + d_{1m}f_m + \varepsilon_1$$
$$Y_2 = d_{21}f_1 + d_{22}f_2 + \ldots + d_{2m}f_m + \varepsilon_2$$
$$\vdots$$
$$Y_k = d_{k1}f_1 + d_{k2}f_2 + \ldots + d_{km}f_m + \varepsilon_k$$

The general FA model is

$$Y_i = d_{i1}f_1 + d_{i2}f_2 + \ldots + d_{im}f_m + \varepsilon_i$$

The number of common factors f_i should be less than the original variables Y_i to meet the objective of FA as a data reduction technique.

The coefficients d_{ij} are called loadings showing the importance (contribution) of each factor to the observable variable or how much the observable variable depends on the $f's$.

Common factors $f_1, f_2, \ldots, f_m$ are uncorrelated with zero mean and unit variance. Furthermore, the specific factors have zero mean and are uncorrelated with common factors.

Easy Statistics for Food Science with R. DOI: https://doi.org/10.1016/B978-0-12-814262-2.00009-1
143

The variance of Y_i is

$$Var(Y_i) = d_{i1}^2 var(f_1) + d_{i2}^2 var(f_i) + \ldots + d_{im}^2 var(f_m) + var(\varepsilon_i)$$

$$Var(Y_i) = d_{i1}^2 + d_{i2}^2 + \ldots + d_{im}^2 + var(\varepsilon_i)$$

It can be observed that the variance of Y_i is partitioned into two parts: one part due to common factors called communality (common variance)

$$d_{i1}^2 + d_{i2}^2 + \ldots + d_{im}^2$$

and another part called the specificity of Y_i (specific variance or residual variance) $var(\varepsilon_i)$. Communalities should be examined to have an idea on how much of the variance in each of the original variables is explained by the extracted factors.

Note:
- The correlation between Y_i and Y_j is

$$r_{ij} = d_{i1}d_{j1} + d_{i2}d_{j2} + \ldots + d_{im}d_{jm}$$

9.3   PROCEDURE FOR FACTOR ANALYSIS

The arrangement of the data for a FA is similar to the arrangement of data for a principal component analysis given in Table 8.1. The procedure for a FA consists of three steps:

- The first step in a FA is to estimate the loadings and communalities. There are many methods to estimate the loadings and communalities such as principal component method, maximum likelihood, principal factor, image factoring, Alpha, and others. The most common method used for extracting the factors is the principal component method.
- The second step is to rotate the factors. Rotation of the factors is used to reduce the contributions of less significant variables and to choose the values of d_{ij}; a value close to zero indicates that the variable does not correlate with this factor, and a value far from zero indicates that the variable is highly correlated with this factor. The rotation can be either orthogonal or oblique, depending on the method used for rotation. There are other methods for rotation, including varimax or orthogonal rotation, oblique rotation, and quartimax.
- Finally, we calculate the factor scores (the values of the factors for each of n samples (observations)) to be used for other purposes, such as to draw a bar chart or scatter diagram or further analysis. Factor scores can be used to study the behavior of the samples (observations).

Statistical packages give options to choose a method for extraction, methods for rotation, and the number of factors to be retained.

9.3.1   Principal Component Method

The principal component method is generally used to extract factors from a correlation or covariance matrix. This method involves the following steps:

- First, calculate the sample covariance matrix (correlation matrix) for the k selected variables to calculate the eigenvalues, and then find the corresponding eigenvectors for each eigenvalue. The principal components will be of the form

$$Z_1 = a_{11}Y_1 + a_{12}Y_2 + a_{13}Y_3 + \ldots + a_{1k}Y_k$$
$$Z_2 = a_{21}Y_1 + a_{22}Y_2 + a_{23}Y_3 + \ldots + a_{2k}Y_k$$
$$\vdots$$
$$Z_k = a_{k1}Y_1 + a_{k2}Y_2 + a_{k3}Y_3 + \ldots + a_{kk}Y_k$$

These are unrotated factors.

- The second step is to find the inverse relationship between Y and Z.

$$Y_1 = a_{11}Z_1 + a_{21}Z_2 + a_{31}Z_3 + \ldots + a_{k1}Z_k$$
$$Y_2 = a_{12}Z_1 + a_{22}Z_2 + a_{32}Z_3 + \ldots + a_{k2}Z_k$$
$$\vdots$$
$$Y_k = a_{1k}Z_1 + a_{2k}Z_2 + a_{3k}Z_3 + \ldots + a_{kk}Z_k$$

- Choose the number of principal components to be retained. The criteria used for principal component analysis can be used for FA. (Only principal components that substantially contribute to explaining the total variance should be kept.) In general, it can be said that only m factors should be kept in the model, as shown below:

$$Y_1 = a_{11}Z_1 + a_{21}Z_2 + a_{31}Z_3 + \ldots + a_{m1}Z_m + e_1$$
$$Y_1 = a_{12}Z_1 + a_{22}Z_2 + a_{32}Z_3 + \ldots + a_{m2}Z_m + e_2$$
$$\vdots$$
$$Y_k = a_{1k}Z_1 + a_{2k}Z_2 + a_{3k}Z_3 + \ldots + a_{mk}Z_m + e_k$$

- The eigenvectors should be normalized to scale the principal components $Z_1, Z_2, \ldots, Z_m$ and obtain the unit variance. This can be obtained easily by dividing Z_i by its standard deviation (the variance is the eigenvalue of each component λ_i); thus, each component is divided by $\sqrt{\lambda_i}$. The factors are still not rotated.

$$Y_1 = g_{11}F_1 + g_{12}F_2 + g_{13}F_3 + \ldots + g_{1m}F_m + e_1$$
$$Y_2 = g_{21}F_1 + g_{22}F_2 + g_{23}F_3 + \ldots + g_{2m}F_m + e_1$$
$$\vdots$$
$$Y_k = g_{k1}F_1 + g_{k2}F_2 + g_{k3}F_3 + \ldots + g_{km}F_m + e_k$$

where $F_i = \frac{Z_i}{\sqrt{\lambda_i}}$ and $g_{ij} = a_{ij}\sqrt{\lambda_i}$

- Rotate the factors using any method that gives acceptable results. The rotated factor model is

$$Y_1 = d_{11}f_1 + d_{12}f_2 + d_{13}f_3 + \ldots + d_{1m}f_m + e_1$$
$$Y_2 = d_{21}f_1 + d_{22}f_2 + d_{23}f_3 + \ldots + d_{2m}f_m + e_1$$
$$\vdots$$
$$Y_k = d_{k1}f_1 + d_{k2}f_2 + d_{k3}f_3 + \ldots + d_{km}f + e_k$$

- The scores for each sample (observation) can be obtained by expressing the rotated factors as a linear combination of the Y values. We use the formula in (9.1) to find the scores.

$$f = (D'D)^{-1}D'Y \tag{9.1}$$

where f represents the rotated factors, Y indicates the standardized Y values, and D is the matrix of factor loadings after rotation.

9.4 FACTOR ANALYSIS IN R

FA can be conducted using the built-in functions provided by the R statistical package. R provides several functions to perform FA, as shown below. Two packages need to be installed from the R library: `psych` and `GPArotation`.

1. The function `principal()` can be used to conduct FA. This function is available in the `psych` packages.

```
principal (data frame, nfactors = ?, rotate = "varimax", covar = FALSE)
```

R uses a correlation matrix by default regardless of whether `covar` is included. If a covariance matrix is desired, then set `covar = TRUE`. The rotation options include `varimax`, `quatimax`, `promax`, `oblimin`, `simplimax`, `cluster`, and `none`.

2. The function `factanal()` can also be used to perform FA. This function is available in the `stats` package.

```
factanal (data frame, factors = Number of factor)
```

R usually uses varimax rotation by default; you can use other options such as Promax and none.

```
factanal (data frame, factors = ?, rotation = "varimax")
```

3. The function `factor.pa()` in the `psych` package offers some factor-analysis-related functions.

```
factor.pa (data, nfactors = ?, rotation = "varimax")
```

The data can be a raw data matrix or a covariance matrix. The rotation can be "varimax" or "promax."

4. The function `$loadings` is used to produce the loadings for the components.

```
Name of the FA $ loadings
```

5. The function `$scores` is used to produce the scores for all of the observations.

```
Name of the FA $ scores
```

6. The `nFactors` package offers additional functions to help one select the number of factors. A crucial decision in exploratory FA is how many factors to extract.
7. The package `FactoMineR` offers some additional functions for exploratory FA, such as the use of both qualitative and quantitative variables.

Example 9.1: Mineral and heavy metal contents in bananas—The levels of major elements (sodium (Na), potassium (K), calcium (Ca), magnesium (Mg)), minor elements (iron (Fe), copper (Cu), zinc (Zn), manganese (Mn)), and one heavy metal (lead (Pb)) in Cavendish banana flour and Dream banana flour have been determined. Because they are two of the most common banana varieties, Cavendish (M. paradisiaca L. cvcavendshii) and Dream (M. acuminatacolla. AAA cv "Berangan") bananas were purchased from 11 markets around Penang, Malaysia. The data obtained from 22 different samples are given in Table 9.1.

The `psych` package and `GPArotation` package should be installed from the R library before implementing FA.

```
> install.packages("psych")
> library(psych)
> install.packages("GPArotation")
> library(GPArotation)
```

The correlation matrix between different mineral and heavy metal contents in bananas was calculated using built-in functions in R. The data had been stored in the `.CSV` (`Example9_1`) format and loaded into the R environment to perform the analysis. The function `corr.test()` was used to calculate the correlations between different variables and to test the significance.

TABLE 9.1 The Data for Mineral and Heavy Metal Contents in Bananas (mg/100g)

	Na	K	Ca	Mg	Cu	Fe	Mn	Zn	pb
Cavendish	88.605	680.159	14.441	94.121	0.301	1.347	0.744	0.642	0.663
Cavendish	109.913	646.141	12.981	93.074	0.370	1.144	0.543	0.859	0.000
Cavendish	129.278	565.939	15.858	91.067	0.435	1.448	0.473	0.641	0.000
Cavendish	108.295	543.273	15.898	98.129	0.467	1.238	0.607	0.852	0.661
Cavendish	74.376	629.635	15.886	94.505	0.394	1.016	0.531	0.734	0.650
Cavendish	84.774	703.956	11.380	101.565	0.497	1.229	0.602	0.951	0.000
Cavendish	47.650	664.673	12.905	95.783	0.462	1.226	0.601	0.844	0.000
Cavendish	62.077	756.601	15.591	102.575	0.496	1.124	0.467	0.739	0.687
Cavendish	51.650	719.111	15.739	106.884	0.360	1.416	0.528	1.044	0.000
Cavendish	42.727	803.976	14.262	104.966	0.394	1.117	0.597	1.048	0.000
Cavendish	43.742	763.351	14.334	98.624	0.492	1.014	0.530	0.837	0.000
Dream	4.155	771.064	14.468	77.990	0.397	1.023	0.535	0.634	0.000
Dream	3.854	714.435	10.103	74.990	0.362	1.119	0.731	0.840	0.977
Dream	2.748	796.965	8.683	83.068	0.434	0.827	0.540	0.853	0.000
Dream	2.463	712.350	8.662	67.811	0.331	1.025	0.535	0.741	0.000
Dream	3.035	686.776	8.648	75.080	0.266	1.234	0.604	0.637	0.658
Dream	2.465	711.157	10.020	69.831	0.336	1.039	0.746	0.643	0.997
Dream	2.904	690.398	11.571	70.376	0.360	1.013	0.596	0.837	0.000
Dream	3.770	756.973	11.360	83.905	0.365	1.027	0.604	0.848	0.000
Dream	3.243	791.091	15.482	73.224	0.430	0.921	0.535	0.634	0.000
Dream	3.255	589.326	12.622	81.140	0.369	1.142	0.746	0.857	0.000
Dream	2.330	474.598	16.611	72.194	0.330	1.020	0.600	0.843	0.000

```
>corr.test(Example9_1[,2:10])
Call:corr.test(x = Example9_1[, 2:10])
Correlation matrix
        Na      K     Ca     Mg     Cu     Fe     Mn     Zn     pb
Na    1.00  -0.35   0.50   0.71   0.40   0.64  -0.27   0.08   0.06
K    -0.35   1.00  -0.34   0.07   0.16  -0.38  -0.16   0.05  -0.08
Ca    0.50  -0.34   1.00   0.50   0.34   0.29  -0.34   0.05  -0.12
Mg    0.71   0.07   0.50   1.00   0.55   0.50  -0.29   0.51  -0.08
Cu    0.40   0.16   0.34   0.55   1.00  -0.06  -0.46   0.23  -0.22
Fe    0.64  -0.38   0.29   0.50  -0.06   1.00   0.07   0.09   0.12
Mn   -0.27  -0.16  -0.34  -0.29  -0.46   0.07   1.00  -0.01   0.44
Zn    0.08   0.05   0.05   0.51   0.23   0.09  -0.01   1.00  -0.38
pb    0.06  -0.08  -0.12  -0.08  -0.22   0.12   0.44  -0.38   1.00
Sample Size
[1] 22
```

(Continued)

(Continued)

Probability values (Entries above the diagonal are adjusted for multiple tests.)

	Na	K	Ca	Mg	Cu	Fe	Mn	Zn	pb
Na	0.00	1.00	0.55	0.01	1.00	0.04	1.00	1.00	1
K	0.11	0.00	1.00	1.00	1.00	1.00	1.00	1.00	1
Ca	0.02	0.12	0.00	0.57	1.00	1.00	1.00	1.00	1
Mg	0.00	0.77	0.02	0.00	0.28	0.55	1.00	0.54	1
Cu	0.07	0.49	0.12	0.01	0.00	1.00	0.88	1.00	1
Fe	0.00	0.08	0.19	0.02	0.80	0.00	1.00	1.00	1
Mn	0.22	0.48	0.13	0.18	0.03	0.76	0.00	1.00	1
Zn	0.73	0.84	0.83	0.02	0.30	0.69	0.98	0.00	1
pb	0.81	0.73	0.59	0.71	0.32	0.60	0.04	0.08	0

To see confidence intervals of the correlations, print with the short = FALSE option

The command to calculate the correlation coefficients among the various variables includes a call to extract the data placed in columns 2−10 of the file Example9_1. Then, we run the command to calculate the correlations between all of the possible combinations of two variables for the data in columns 2−10. The correlation matrix for the mineral and heavy metal contents in bananas was computed. Although a strong positive correlation was revealed between Na with Ca, Mg, and Fe, K did not correlate with the other parameters. Furthermore, Mg was positively correlated with Ca, Cu, Fe, and Zn. A strong negative correlation was exhibited between Mn and Cu, and Mn was positively correlated with Pb.

A FA was carried out on the data set (with nine variables). The principal component method was used to extract the factors using a correlation matrix of the selected variables (elements and heavy metals). The function princomp() was used to calculate the square roots of the eigenvalues (standard deviation), the percentage of the variance that is accounted for (proportion of variance), and the cumulative percentage variance for the unrotated factors (cumulative proportion). Then, we saved the results in pca.

```
> PCA <- princomp(Example9_1[,2:10],cor = TRUE) # principal components method
> summary(PCA)
Importance of components:
                        Comp.1      Comp.2      Comp.3      Comp.4      Comp.5
Standard deviation    1.7791449   1.3997716   1.0827313   1.0120092   0.80568889
Proportion of Variance 0.3517063  0.2177067   0.1302563   0.1137959   0.07212607
Cumulative Proportion  0.3517063  0.5694130   0.6996694   0.8134652   0.88559128
                        Comp.6      Comp.7      Comp.8      Comp.9
Standard deviation    0.68344402  0.5614106   0.43286078  0.245015298
Proportion of Variance 0.05189953 0.0350202   0.02081872  0.006670277
Cumulative Proportion  0.93749080 0.9725110   0.99332972  1.000000000
```

FA resulted in only four factors with eigenvalues greater than one ($(1.779)^2$, $(1.399)^2$, $(1.0827)^2$, and $(1.012)^2$), and the variance accounted for by the first four factors consists of 81.346% of the total variance in the elements and Pb data. Therefore, these factors represent the selected variables very well. An eigenvalue gives a measure of the significance of the factors; the factors with the highest eigenvalues are the most significant and can explain large variations in the data. A Scree plot for the extracted components was produced using screeplot(), and it is given in Fig. 9.1.

The four unrotated factor loadings are extracted and rounded to two decimals.

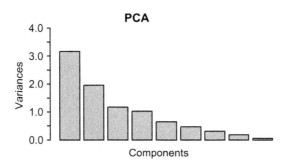

FIGURE 9.1 Scree plot for the eigenvalues extracted from the mineral and heavy metal contents in bananas.

```
round (unclass (PCA $ loadings), digits = 2)
      Comp.1  Comp.2  Comp.3  Comp.4  Comp.5  Comp.6  Comp.7  Comp.8  Comp.9
Na     0.47    0.25    0.08   -0.14    0.15    0.31    0.05    0.64   -0.40
K     -0.12   -0.47   -0.14   -0.57    0.33   -0.39   -0.22    0.10   -0.31
Ca     0.39    0.09    0.33    0.19   -0.37   -0.70   -0.14   -0.02   -0.21
Mg     0.50   -0.04   -0.27   -0.28    0.02   -0.18    0.06    0.16    0.74
Cu     0.35   -0.33    0.13   -0.29   -0.42    0.47   -0.35   -0.36   -0.08
Fe     0.31    0.44   -0.20    0.01    0.52   -0.02   -0.25   -0.55   -0.14
Mn    -0.27    0.39   -0.47   -0.07   -0.37   -0.03   -0.60    0.24    0.00
Zn     0.21   -0.23   -0.72    0.23   -0.23   -0.03    0.38   -0.10   -0.35
pb    -0.14    0.45    0.05   -0.63   -0.29   -0.07    0.48   -0.22   -0.11
```

The function `principal()` was used to perform FA, including a varimax rotation with a Kaiser normalization to produce the factor model. The variance and cumulative variance explained by each factor after rotation, the factor loadings, the communalities, and the specific variance were calculated using R commands and built-in functions.

```
> Factor <- principal (Example9_1[2:10], nfactors = 4, rotate = "varimax")
> print (Factor, digits = 2)
Principal Components Analysis
Call: principal(r = Example9_1[2:10], nfactors = 4, rotate = "varimax")
Standardized loadings (pattern matrix) based upon correlation matrix
      PC1    PC4    PC3    PC2     h2     u2    com
Na    0.85   0.25  -0.07  -0.24   0.85  0.146  1.4
K    -0.18   0.08   0.05   0.89   0.83  0.166  1.1
Ca    0.47   0.53  -0.04  -0.41   0.67  0.327  2.9
Mg    0.87   0.25   0.32   0.18   0.95  0.054  1.5
Cu    0.41   0.63   0.11   0.37   0.71  0.285  2.5
Fe    0.73  -0.21   0.04  -0.40   0.74  0.257  1.8
Mn   -0.06  -0.88  -0.03  -0.07   0.79  0.214  1.0
Zn    0.26  -0.08   0.91   0.13   0.91  0.090  1.2
pb    0.28  -0.54  -0.68   0.15   0.86  0.141  2.4

                        PC1    PC4    PC3    PC2
SS loadings             2.59   1.93   1.41   1.39
Proportion Var          0.29   0.21   0.16   0.15
Cumulative Var          0.29   0.50   0.66   0.81
Proportion Explained    0.35   0.26   0.19   0.19
Cumulative Proportion   0.35   0.62   0.81   1.00
```

(Continued)

(Continued)
Mean item complexity = 1.8
Test of the hypothesis that 4 components are sufficient.
The root mean square of the residuals (RMSR) is 0.07
with the empirical chi square 8.76 with prob < 0.19
Fit based upon off diagonal values = 0.95

The first line of the output shows the call to execute FA using the function `principal()` and then store the results in a new place called `Factor`; next, there is a command to print the result. Standardized loadings based on a correlation matrix appear as `PC1`, `PC4`, `PC3`, and `PC2`, then there are two columns for the communalities (`h2`) and specific variance (`u2` = 1—communality). These communalities usually indicate how much of the variance in the mineral and heavy metal contents in bananas is explained by the extracted factors. The last results of the `principal()` function are the proportion of variance (`proportion Var`) explained by each factor and the cumulative variance (`Cumulative Var`).

The new factors (after rotation) are considerably more sensible than before the rotation due to the reduction in the contributions of the less significant variables (parameters). The improvement is clear from the amount of variance explained by each factor; for instance, the amount of variance explained by the first factor before the rotation was 35%, but after the rotation, this figure changed to 29%. The change is due to the reduction of the effect of the factor that has a small contribution with the first factor.

Hypothesis testing using the chi-square test showed that four factors are sufficient to explain or capture most of the variance in the data (more than 81%). Notice that most of the variables associated with each factor are well defined and contribute slightly to the other factors.

The parameter loadings that were greater than 0.60 were considered to have a strong correlation within each factor. These loadings illustrate that most of the variables associated with each factor are well defined and contribute very little to the other factors, which assists the interpretation of the results. The four factors are presented in Eqs. (9.2–9.5).

$$F_1 = 0.85\text{Na} - 0.18k + 0.47\text{Ca} + 0.87\text{Mg} + 0.41\text{Cu} + 0.73\text{Fe} - 0.06\text{Mn} + 0.26\text{Zn} + 0.28\text{pb} \tag{9.2}$$

$$F_2 = 0.25\text{Na} + 0.08k + 0.53\text{Ca} + 0.25\text{Mg} + 0.63\text{Cu} - 0.21\text{Fe} - 0.88\text{Mn} - 0.08\text{Zn} - 0.54\text{pb} \tag{9.3}$$

$$F_3 = -0.07\text{Na} + 0.05k - 0.04\text{Ca} + 0.32\text{Mg} + 0.11\text{Cu} + 0.04\text{Fe} - 0.03\text{Mn} + 0.91\text{Zn} - 0.68\text{pb} \tag{9.4}$$

$$F_4 = -0.24\text{Na} + 0.89k - 0.41\text{Ca} + 0.18\text{Mg} + 0.37\text{Cu} - 0.40\text{Fe} - 0.07\text{Mn} + 0.13\text{Zn} + 0.15\text{pb} \tag{9.5}$$

The first factor explained 29% of the total variance and was positively correlated with Mg, Na, and Fe, and the second factor explained 21% of the total variance and was negatively correlated with Mn and positively correlated with Cu. In contrast, the third factor accounted for 16% of the total variance and was correlated positively with Zn and negatively with Pb; lastly, the fourth factor accounted for 15% of the total variance and was positively correlated with one parameter (K). Many factors may affect the variation in the mineral and heavy metal contents of banana flour, such as the variety of banana used and the bananas' stage of ripeness, as well as the soil type, soil condition, fertilization, irrigation, and weather. Because ripened bananas were used in this study (they were obtained at the point of sale), the variation could be attributed to the variation in the concentration of elements and Pb with the varieties, agricultural practices, and locations of plantations. The communalities were produced by the function `principal()` and are calculated from the sums of the squares of the loadings for the extracted factors. For instance, the first communality is calculated as $(0.85^2 + 0.25^2 + (-0.07)^2 + (-0.24)^2 = 0.85)$. The communalities can be used to assess the model; the values close to one indicate that the model explains most of the variation for the selected variables. The results can be verified by checking the communalities for the mineral and heavy metal contents in bananas. Indeed, these communalities are high, which indicates that the model captured most of the variance in the data. The factor score for each observation was calculated using the function `$scores`.

```
> round (Factor $ scores, digits = 2)
          PC1     PC4     PC3     PC2
   [1,]   1.22   -1.49   -1.15   -0.54
   [2,]   0.46    0.31    0.50   -0.60
   [3,]   1.08    1.36   -0.90   -1.74
   [4,]   1.60    0.01   -0.40   -0.73
   [5,]   0.52    0.65   -1.10   -0.25
   [6,]   1.11   -0.04    1.14    0.75
   [7,]   0.52    0.16    0.62    0.02
   [8,]   1.15    1.20   -1.26    1.30
   [9,]   1.16   -0.17    1.75   -0.28
  [10,]   0.64   -0.23    1.74    1.02
  [11,]   0.25    1.15    0.31    1.05
  [12,]  -0.98    1.11   -0.77    0.21
  [13,]   0.01   -1.97   -0.44    0.82
  [14,]  -1.12    0.44    0.57    1.63
  [15,]  -1.57    0.03    0.05   -0.06
  [16,]  -0.59   -1.30   -1.08   -0.18
  [17,]  -0.48   -1.72   -1.56    0.61
  [18,]  -1.25   -0.06    0.55   -0.30
  [19,]  -0.75   -0.21    0.71    0.50
  [20,]  -1.21    1.51   -0.93    0.44
  [21,]  -0.51   -1.07    1.03   -1.06
  [22,]  -1.26    0.34    0.61   -2.61
```

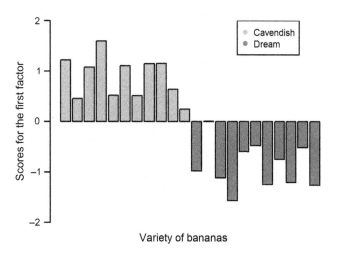

FIGURE 9.2 Factor scores for the first factor by the variety of banana.

The relationship between the factor scores and the samples was studied to gain a deeper understanding of the behavior of the concentrations of elements and Pb in selected fruits. Bar charts for the factor scores were produced using the barplot() function in R (see the appendix for the R code). The factor scores for the first factor are shown in Fig. 9.2. Observe that all of the Cavendish banana fruit samples exhibited a positive contribution to the first factor, whereas all of the Dream banana fruit samples exhibited a negative contribution to this factor. A positive contribution was attributed to the concentrations of Mg, Na, and Fe, whereas a negative contribution was attributed to the concentrations of Mn and K. Thus, the two types of flour were different with regard to these elements.

The scores for the second factor are shown in Fig. 9.3. (see the appendix for the R code.) The positive contribution was mainly due to the concentrations of Cu and Ca, and the negative contribution was primarily due to the

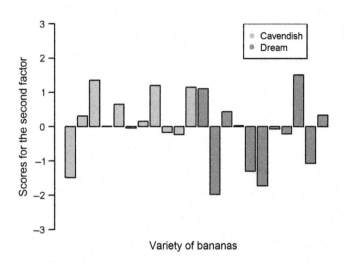

FIGURE 9.3 Factor scores for the second factor by the variety of banana.

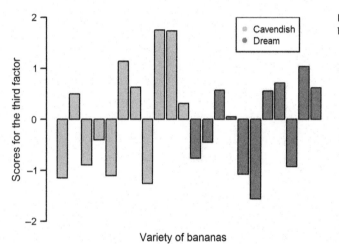

FIGURE 9.4 Factor scores for the third factor by the variety of banana.

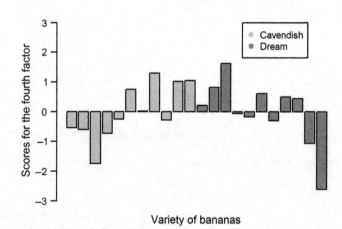

FIGURE 9.5   Factor scores for the fourth factor by the variety of banana.

concentration of Mn in the flour; a smaller effect was observed from Pb and Fe. The concentrations of minerals exhibited different behavior from sample to sample regardless of the types of flour.

The scores for factors 3 and 4 (Figs. 9.4 and 9.5) exhibited positive and negative contributions regardless of the type of flour.

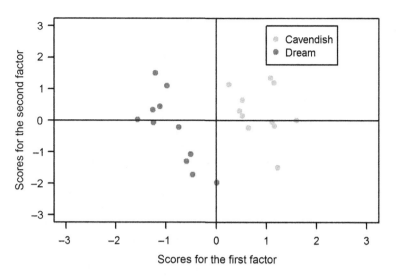

FIGURE 9.6 Factor scores for the first and second factors by the variety of banana.

The scores for the first two factors were plotted together (scatter diagram) and presented in Fig. 9.6. The graph revealed that although the first factor was responsible for separating the two types of banana flour, the second factor represents the fluctuations in the concentrations of the minerals and Pb regardless of the type of flour.

FA has provided a satisfactory picture of the concentration of elements and Pb in banana flour. FA provides valuable information about the source of the variation in the elements and Pb, as more than 81% of the total variance was explained by only four factors.

Example 9.2: Quality characteristics for deep-fried food—Quality characteristics for tapioca deep-fried chips were studied in terms of the fat oxidation, fat content, fatty acid composition, and oil color. Seven different brands of tapioca were selected based on the packaging type (polyethylene zip-lock, laminated gas packaging, and typical transparent polypropylene). The samples were analyzed for their fat oxidation (Peroxide (PV), p-anisidine (PAV), acid value (AV) [mg KOH/g] and thiobarbituric acid (TBA) (mg MAD/kg)), fat content (%), and fatty acid composition (%) (saturated fatty acid (SFA), polyunsaturated fatty acid (PUFA), monounsaturated fatty acid (MUFA)), oil color values (the lightness (L*) value and the redness (a*), yellowness (b*)), and moisture content (%). The data are given in Table 1.4.

The correlation matrix of quality characteristics parameters for tapioca deep-fried chips was calculated. The entries of the correlation matrix were calculated using built-in functions in R. The function corr.test() was used to compute the bivariate correlation between different variables and test the significance of the relationship. The data had been stored as a .CSV (Example9_2) file.

```
>corr.test(Example9_2[,2:13])
Call:corr.test(x = Example9_2[, 2:13])
Correlation matrix
           FC    Mo.     L      a      b     PV    PAV     AV    TBA    SAF   MUFA   PUFA
FC       1.00   0.72   0.18   0.22   0.86   0.10  -0.34  -0.29  -0.32   0.02   0.11   0.41
Mo.      0.72   1.00   0.00   0.10   0.34   0.16  -0.43   0.02  -0.06  -0.10  -0.07   0.08
L        0.18   0.00   1.00  -0.78   0.44  -0.79   0.06  -0.93  -0.67   0.53   0.72   0.77
a        0.22   0.10  -0.78   1.00   0.03   0.72  -0.43   0.56   0.22  -0.25  -0.56  -0.33
b        0.86   0.34   0.44   0.03   1.00  -0.24  -0.09  -0.56  -0.60   0.06   0.26   0.56
PV       0.10   0.16  -0.79   0.72  -0.24   1.00  -0.32   0.69   0.77  -0.06  -0.41  -0.38
PAV     -0.34  -0.43   0.06  -0.43  -0.09  -0.32   1.00   0.18   0.13  -0.53  -0.30  -0.47
AV      -0.29   0.02  -0.93   0.56  -0.56   0.69   0.18   1.00   0.74  -0.63  -0.83  -0.92
TBA     -0.32  -0.06  -0.67   0.22  -0.60   0.77   0.13   0.74   1.00  -0.16  -0.30  -0.57
SAF      0.02  -0.10   0.53  -0.25   0.06  -0.06  -0.53  -0.63  -0.16   1.00   0.72   0.78
MUFA     0.11  -0.07   0.72  -0.56   0.26  -0.41  -0.30  -0.83  -0.30   0.72   1.00   0.87
PUFA     0.41   0.08   0.77  -0.33   0.56  -0.38  -0.47  -0.92  -0.57   0.78   0.87   1.00
Sample Size
[1] 21
```

(Continued)

(Continued)

Probability values (Entries above the diagonal are adjusted for multiple tests.)

	FC	Mo.	L	a	b	PV	PAV	AV	TBA	SAF	MUFA	PUFA
FC	0.00	0.01	1.00	1.00	0.00	1.00	1.00	1.00	1.00	1.00	1.00	1.00
Mo.	0.00	0.00	1.00	1.00	1.00	1.00	1.00	1.00	1.00	1.00	1.00	1.00
L	0.44	0.98	0.00	0.00	1.00	0.00	1.00	0.00	0.04	0.56	0.01	0.00
a	0.34	0.67	0.00	0.00	1.00	0.01	1.00	0.38	1.00	1.00	0.38	1.00
b	0.00	0.13	0.04	0.89	0.00	1.00	1.00	0.37	0.19	1.00	1.00	0.38
PV	0.65	0.49	0.00	0.00	0.29	0.00	1.00	0.03	0.00	1.00	1.00	1.00
PAV	0.13	0.05	0.79	0.05	0.69	0.16	0.00	1.00	1.00	0.56	1.00	1.00
AV	0.21	0.92	0.00	0.01	0.01	0.00	0.43	0.00	0.01	0.11	0.00	0.00
TBA	0.16	0.79	0.00	0.35	0.00	0.00	0.58	0.00	0.00	1.00	1.00	0.34
SAF	0.94	0.65	0.01	0.27	0.80	0.81	0.01	0.00	0.48	0.00	0.01	0.00
MUFA	0.65	0.77	0.00	0.01	0.26	0.07	0.19	0.00	0.19	0.00	0.00	0.00
PUFA	0.07	0.73	0.00	0.14	0.01	0.09	0.03	0.00	0.01	0.00	0.00	0.00

To see confidence intervals of the correlations, print with the short = FALSE option

The correlation matrix of the quality characteristics parameters for tapioca showed that most of the variables (parameters) had a significant bivariate relationship with the other variables.

A FA was conducted on the data set (12 variables) to identify the sources of variation in the data and to identify the parameters responsible for the fluctuations in the data. The principal component method was used to extract the factors from the correlation matrix. In addition, the function `princomp()` was used to produce the square roots of the eigenvalues (Standard deviation), the percentage variance accounted for (Proportion of variance), and the cumulative percentage variance (Cumulative proportion) for the unrotated factors.

```
> PCA <- princomp (Example9_2, cor = TRUE)
> summary (PCA)
Importance of components:
                           Comp.1        Comp.2        Comp.3        Comp.4        Comp.5
Standard deviation        2.3709498     1.6543476     1.3708471     0.93189688    0.80011349
Proportion of Variance    0.4684503     0.2280722     0.1566018     0.07236932    0.05334847
Cumulative Proportion     0.4684503     0.6965224     0.8531243     0.92549357    0.97884204
                           Comp.6        Comp.7        Comp.8        Comp.9
Standard deviation        0.47506295    0.133971127   0.0705214392  0.0508044582
Proportion of Variance    0.01880707    0.001495689   0.0004144394  0.0002150911
Cumulative Proportion     0.99764911    0.999144797   0.9995592360  0.9997743271
                           Comp.10       Comp.11       Comp.12
Standard deviation        0.0433358873  2.691954e-02  1.026716e-02
Proportion of Variance    0.0001564999  6.038847e-05  8.784556e-06
Cumulative Proportion     0.9999308270  9.999912e-01  1.000000e+00
```

The results of `princomp()` showed that only three eigenvalues are greater than one: 2.3709, 1.6543, and 1.3708. (This criterion is employed because the correlation matrix was used to extract the factors.) Thus, only three factors can be used to describe the relationship between the selected variables and the factors. The variance accounted for by the first three factors is 85.31% of the total variance, and therefore, these three factors represent the 12 variables very well. The proportion of the total variance for each factor is the variance (eigenvalue), which is calculated by dividing by the number of variables (here, 12). For instance, the proportion of the variance explained by the first component is $(((2.3709)^2/12) = 0.468)$. The Scree plot for the calculated eigenvalues for tapioca deep-fried-chips data is given in Fig. 9.7.

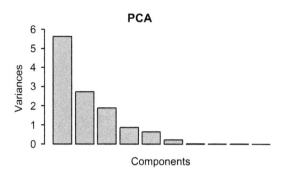

FIGURE 9.7 Scree plot for the eigenvalues extracted from different brands of tapioca.

The loadings for unrotated factors were produced using the function $loadings, and the results were rounded to two decimals.

```
> round(unclass(PCA$loadings),digits=2)
      Comp.1  Comp.2  Comp.3  Comp.4  Comp.5  Comp.6  Comp.7  Comp.8  Comp.9  Comp.10 Comp.11 Comp.12
FC    -0.14   -0.50   0.26    -0.16   -0.25   0.04    -0.09   0.35    0.10    0.46    0.45    0.13
Mo    -0.03   -0.43   0.18    -0.60   0.42    0.06    -0.10   -0.34   -0.05   -0.34   0.02    0.02
L     -0.40   0.15    0.08    -0.14   0.01    0.31    0.36    0.14    -0.04   -0.01   -0.29   0.68
a     0.24    -0.38   -0.07   0.54    -0.02   -0.07   -0.10   -0.23   0.20    -0.29   0.14    0.54
b     -0.24   -0.32   0.36    0.16    -0.44   -0.01   -0.13   0.10    0.01    -0.32   -0.54   -0.27
PV    0.29    -0.31   -0.31   -0.09   -0.36   0.17    0.26    -0.27   -0.60   0.22    -0.15   0.01
PAV   0.09    0.42    0.41    -0.13   -0.48   0.18    -0.12   -0.31   -0.13   -0.26   0.41    0.10
AV    0.42    -0.03   0.01    -0.09   0.05    0.03    0.20    0.69    -0.23   -0.48   0.13    0.01
TBA   0.31    0.03    -0.30   -0.46   -0.40   -0.11   0.03    0.00    0.62    -0.03   -0.19   0.10
SAF   -0.26   -0.04   -0.53   0.01    -0.07   0.61    -0.45   0.12    0.02    -0.20   0.14    -0.09
MUFA  -0.35   0.05    -0.30   -0.17   -0.17   -0.66   -0.28   0.09    -0.34   -0.16   0.08    0.24
PUFA  -0.39   -0.13   -0.20   0.08    -0.13   -0.08   0.65    -0.13   0.18    -0.28   0.37    -0.27
```

Rotation of the factors can reduce the contributions of the less significant variables and provide simpler factors. A varimax rotation with Kaiser normalization was conducted to produce the factor model for the quality characteristics of deep-fried-food parameters. The variance and cumulative variance explained by each factor after rotation, the loadings for the rotated factors with the communalities, and the specific variance were calculated using the function principal().

```
> Factor <- principal (Example9_2, nfactors = 3, rotate = "varimax")
> Factor
Principal Components Analysis
Call: principal(r = Example9_2,nfactors = 3,rotate = "varimax")
Standardized loadings (pattern matrix) based upon correlation matrix

      PC1     PC3     PC2     h2      u2      com
FC    0.00    0.11    0.96    0.94    0.062   1.0
Mo    0.19    0.01    0.73    0.57    0.425   1.1
L     -0.87   0.44    0.11    0.96    0.041   1.5
a     0.79    -0.14   0.32    0.74    0.257   1.4
b     -0.38   0.09    0.84    0.85    0.148   1.4
PV    0.95    0.01    0.04    0.90    0.097   1.0
PAV   -0.44   -0.70   -0.38   0.83    0.167   2.3
```

(Continued)

(Continued)

AV	0.75	-0.61	-0.23	0.99	0.012	2.1
TBA	0.70	-0.14	-0.44	0.71	0.294	1.8
SAF	-0.10	0.94	-0.11	0.91	0.090	1.1
MUFA	-0.46	0.80	-0.03	0.86	0.145	1.6
PUFA	-0.45	0.82	0.32	0.98	0.025	1.9

	RC1	RC3	RC2
SS loadings	4.13	3.32	2.79
Proportion Var	0.34	0.28	0.23
Cumulative Var	0.34	0.62	0.85
Proportion Explained	0.40	0.32	0.27
Cumulative Proportion	0.40	0.73	1.00

Mean item complexity = 1.5
Test of the hypothesis that 3 components are sufficient.

The root mean square of the residuals (RMSR) is 0.08
with the empirical chi square 16.54 with prob< 0.99

Fit based upon off diagonal values = 0.98

The new factors after rotation are considerably more sensible than before the rotation (principal component), as each variable is associated with one factor even though it appears with others that have a low coefficient. The communalities are obtained from the sums of the squares of the loadings for the three factors; for instance, the first communality (h2) is ($(0.0)^2 + 0.11^2 + 0.96^2 = 0.94$). Most of the communalities are high, which indicates that most of the variance for the variables is accounted for by three factors. The specific variance is equal to $u2 = 1$—communality. Large factor loadings show how the variables are related to the factors.

Hypothesis testing using the chi-square test showed that the three factors are sufficient to describe the relationship and reveal the hidden information.

Most of the variables associated with each factor are well defined and contribute slightly to the other factors, which assists the interpretation of the results and the identification of the sources of difference in tapioca deep-fried food.

After the rotation, the factors provided better results since the relationship between the factors and the variables was clearer than before.

The three factors are presented in equation form in (9.6−9.8).

$$F_1 = 0.0FC + 0.19MO. - 0.87L^* + 0.79a^* - 0.38b^* + 0.95PV - 0.44PAV + 0.75AV$$
$$+ 0.75TBA - 0.10SAF - 0.46MUFA - 0.45PUFA \tag{9.6}$$

$$F_2 = 0.11FC + 0.01MO. + 0.44L^* - 0.14a^* + 0.09b^* + 0.01PV - 0.70PAV - 0.61AV$$
$$- 0.14TBA + 0.94SAF + 0.80MUFA + 0.82PUFA \tag{9.7}$$

$$F_3 = 0.96FC + 0.73MO. + 0.11L^* + 0.32a^* + 0.84b^* + 0.04PV - 0.38PAV - 0.23AV$$
$$- 0.44mTBA - 0.11SAF - 0.03MUFA + 0.32PUFA \tag{9.8}$$

Although the results are now ready for interpretation, it is better to label the factors based on the high loading associated with each factor. The first factor accounted for 34% of the total variance and was positively correlated with PV, AV, TBA, and a*, but it was negatively correlated with L*, as shown in the results. This factor may be labeled the quality characteristic of the extracted oil factor because PV, AV, and TBA were used to monitor the deterioration of the extracted oil from fried tapioca chips. However, the second factor accounted for 28% of the total variance and was positively correlated with saturated, monounsaturated, and polyunsaturated fatty acids and negatively with PAV. The third factor accounted for 23% of the total variance and was positively correlated with the fat content, moisture content, and b* value. Many causes may contribute to the variations in the extracted oil of fried tapioca chips. The possibilities include the quality deterioration of tapioca for processing, frying practices, packaging strategies, storage, and distribution condition, as the quality deterioration of tapioca used during processing is critical due to the reduction of the product shelf life. Furthermore, an advanced packaging system has advantages for the retention of the product's color, flavor, and shelf life.

The relationship between the factor scores and the samples from different brands were studied to elucidate the behavior of the selected parameters from various brands. The factor scores for tapioca deep-fried-chips data can be produced using R commands. The function $scores was used to generate factor scores for the tapioca deep-fried-chips data.

```
> round (Factor $ scores, digits = 2)
          PC1      PC3      PC2
[1,]    -0.95    -0.23    -0.61
[2,]    -0.98    -0.20    -0.71
[3,]    -0.95    -0.18    -0.65
[4,]     1.99    -1.09    -0.79
[5,]     1.96    -1.07    -0.72
[6,]     2.00    -0.90    -0.85
[7,]    -0.55     0.67    -0.77
[8,]    -0.52     0.75    -0.85
[9,]    -0.54     0.79    -0.98
[10,]    0.30     0.43     0.39
[11,]    0.34     0.46     0.42
[12,]    0.28     0.40     0.42
[13,]   -0.38     0.72    -0.78
[14,]   -0.40     0.74    -0.81
[15,]   -0.36     0.73    -0.88
[16,]    0.58     1.08     1.73
[17,]    0.57     1.10     1.76
[18,]    0.57     1.10     1.71
[19,]   -0.96    -1.85     1.07
[20,]   -1.06    -1.80     0.99
[21,]   -0.94    -1.63     0.91
```

R can be used to produce different types of graphical representations, such as bar charts, histograms, scatter diagrams, and other types of graphs. The Barplot() function was used to generate a bar chart for the factor scores of the different factors obtained from FA to visualize the behavior of the selected parameters for different brands. The relationship between the factor scores and the samples of various brands was studied. The factor scores for the first factor for various brands are given in Fig. 9.8.

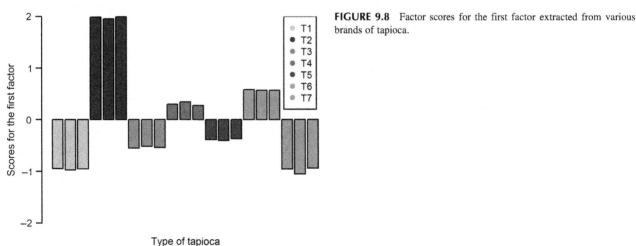

FIGURE 9.8 Factor scores for the first factor extracted from various brands of tapioca.

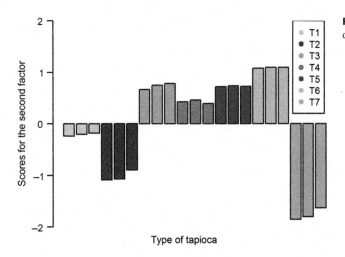

FIGURE 9.9   Factor scores for the second factor extracted from various brands of tapioca.

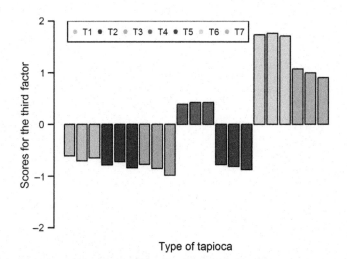

FIGURE 9.10   Factor scores for the third factor extracted from various brands of tapioca.

Although some brands exhibited a positive contribution to this factor, other brands exhibited a negative contribution to it. The positive contribution is attributed to high values of PV, AV, TBA, and a*, whereas the negative contribution is due to the high value of L*. This discovery indicated that the seven brands of tapioca chips were different in terms of the selected parameters. The scores for the second factor are shown in Fig. 9.9.

Whereas the positive contribution was due to the saturated, monounsaturated, and polyunsaturated fatty acids, the negative contribution was due to the PAV. The amount of fatty acid composition exhibited different behavior from sample to sample; the amount of fatty acid depends on the types of oil used and the frying practices. Different behavior was exhibited relative to factor 3, as shown in Fig. 9.10; the positive contribution was due to the fat content, moisture content, and b* value. This correlation indicated that the seven brands of fried tapioca chips were different in terms of these parameters because of the thickness of the chips and the migration of oil into the food structure, which depends on the cooking time and the temperature of the frying oil.

The scores for the first and second factors were plotted in a scatter diagram (Fig. 9.11), which showed that seven different groups can be identified. This result indicates that the seven brands are different in terms of the selected parameters.

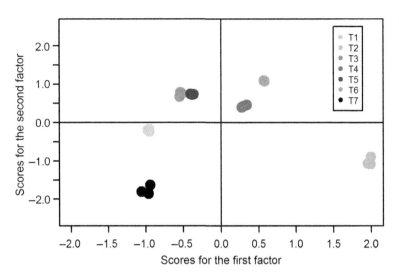

FIGURE 9.11 Factor scores for the first and second factors extracted from various brands of tapioca.

FURTHER READING

Alkarkhi, A. F. M., Saifullah, B. R., & Azhar, M. E. (2009). Application of multivariate statistical techniques for differentiation of ripe banana flour based on the composition of elements. *International Journal of Food Sciences and Nutrition*, *60*, 116–125.

Alkarkhi, A. M., Alqaraquli, W. A. A., Nurl, H., & Yusri, Y. Tapioca. Unpublished work.

Bryan, F. J. M. (1991). *Multivariate statistical methods: A primer*. Great Britain: Chapman & Hall.

Daniel, H. (2013). *Blog Archives, High Resolution Figures in R* [Online]. Available: <https://www.r-bloggers.com/author/daniel-hocking/> Accessed 15.07.17.

Gaston, S. (2012). *5 Functions to do Principal Components Analysis in R* [Online]. Available: <http://gastonsanchez.com/visually-enforced/how-to/2012/06/17/PCA-in-R/> Accessed 15.07.17.

James, H. S. *Exploratory Factor Analysis with R*. Available: http://www.statpower.net/Content/312/R%20Stuff/Exploratory%20Factor%20Analysis%20with%20R.pdf.

Johnson, R. A., & Wichern, D. W. (2002). *Applied multivariate statistical analysis*. New Jersey: Prentice Hall.

R Documentation. *Factor Analysis, Factanal {stats}* [Online]. Available: https://stat.ethz.ch/R-manual/R-devel/library/stats/html/factanal.html.

Rencher, A. C. (2002). *Methods of multivariate analysis*. New York: J. Wiley.

Robert, I. K. (2017). *Quick-R, accessing the power of R - Principal Components and Factor Analysis* [Online]. Available: <http://www.statmethods.net/advstats/factor.html> Accessed 25.08.17.

William, R. (2017a). How To: Use the Psych Package for Factor Analysis and Data Reduction.

William, R. (2017b). Procedures for Psychological, Psychometric, and Personality Research - Package 'psych'.

Chapter 10

Discriminant Analysis and Classification

Learning outcomes

At the end of this chapter, you should be able

- *To describe discriminant analysis (DA).*
- *To describe the classification procedure using a discriminant function (DF).*
- *To know how to apply DA to food science data.*
- *To apply R commands and built-in functions for DA.*
- *To understand and interpret the results of DA.*
- *To report on the useful conclusions.*

10.1 INTRODUCTION

Discriminant analysis (DA) is a multivariate technique used to separate two or more groups of observations (individuals) based on k variables measured on each experimental unit (sample) and find the contribution of each variable in separating the groups. DA works by finding one or more linear combinations of the k selected variables. Furthermore, prediction or allocation of new observations to previously defined groups can be investigated with a linear or quadratic function to assign each individual to one of the predefined groups. For example, this method could be used to separate four types of flour prepared from green and ripe Cavendish bananas based on physicochemical properties (green peel (Gpe), ripe peel (Rpe), green pulp (Gpu), and ripe pulp (Rpu)).DA has been widely used for analyzing food science data to separate different groups.

DA is typically used when the groups are already defined prior to the study.

The end result of DA is a model that can be used for the prediction of group memberships. This model allows us to understand the relationship between the set of selected variables and the observations. Furthermore, this model will enable one to assess the contributions of different variables.

10.2 DESCRIBING DISCRIMINANT ANALYSIS DATA

Suppose there are m populations (groups), and a sample is selected from each population of sizes $n_1, n_2, \ldots, n_m$ with k variables measured from each sample. The arrangement of data for a DA is given in Table 10.1.

The subscript associated with Y_{ijl} in Table 10.1 is explained as follows:

i represents the ith observation $i = 1, 2, 3, \ldots, n_i, j$ represents the population (sample) ($j = 1, 2, 3, \ldots, m$, and l represents the variable measured from each population ($l = 1, 2, 3, \ldots, k$.

The data in Table 10.1 will be used to find a linear combination of the k measured variables to separate the m populations. This linear combination is called the discriminant function (DF).

10.3 DISCRIMINANT FUNCTION

It is useful to determine a function that can be used to separate different groups and to identify the relative contribution for each variable in separating the different groups. We can study the simplest case where there are two groups and k

Easy Statistics for Food Science with R. DOI: https://doi.org/10.1016/B978-0-12-814262-2.00010-8

TABLE 10.1 The General Arrangement of Data for a Discriminant Analysis

Variable individuals	Y_1	Y_2	...	Y_k	Population
1	Y_{111}	Y_{112}	...	Y_{11k}	Population 1
2	Y_{211}	Y_{212}	...	Y_{21k}	
⋮	⋮	⋮	...	⋮	
n_1	Y_{n_111}	Y_{n_112}	...	Y_{n_11k}	
1	Y_{121}	Y_{122}	...	Y_{12k}	Population 2
2	Y_{221}	Y_{222}	...	Y_{22k}	
⋮	⋮	⋮	...	⋮	
n_2	Y_{n_221}	Y_{n_222}	...	Y_{n_22k}	
⋮	⋮	⋮	⋮	⋮	
1	Y_{1m1}	Y_{1m2}	...	Y_{1mk}	Population m
2	Y_{2m1}	Y_{2m2}	...	Y_{2mk}	
⋮	⋮	⋮	...	⋮	
n_m	Y_{n_mm1}	Y_{n_mm2}	...	Y_{n_mmk}	

variables ($Y's$) measured from each group. To separate the two groups, we should determine a linear combination of the variables $Y's$ that maximizes the distance between the two mean vectors. The linear combination that can be used for separation is

$$Z = a_1Y_1 + a_2Y_2 + \ldots + a_kY_k \tag{10.1}$$

This function is called the DF (Canonical DF), and a_i represents the coefficient associated with each variable to be estimated.

In the case of several groups (m) to be compared with k variable measured from each group and n_i observations in the ith group, the DFs are given below.

$$Z = a_{i1}Y_1 + a_{i2}Y_2 + \ldots + a_{ik}Y_k$$

The first DF is

$$Z_1 = a_{11}Y_1 + a_{12}Y_2 + \ldots + a_{1k}Y_k$$

The first discriminant function (DF_1) reflects the differences among the groups as much as possible. The second DF is

$$Z_2 = a_{21}Y_1 + a_{22}Y_2 + \ldots + a_{2k}Y_k$$

The second discriminant function (DF_2) reflects the differences among the groups that are not captured by Z_1. Z_1 and Z_2 are uncorrelated.

The number of linear combinations available for separating different groups is the smaller of k and $m - 1$, says. Furthermore, Z_i is uncorrelated with $Z_1, Z_2, \ldots, Z_{i-1}$.

10.3.1 Procedure for Finding the Discriminant Function

The procedure for finding DF requires finding the coefficients of the DFs (a_i).

The entries of a MANOVA can be used to calculate the a_i values. The procedure for finding a DF consists of the following steps:

- First, calculate the between-groups sum of squares (B) and within the (error) sum of squares (W).
- Second, calculate the eigenvalues (λ) for the matrix $W^{-1}B$, and then calculate the eigenvector associated with each eigenvalue to represent a_i. The eigenvalues can be arranged in order, such that $\lambda_1 > \lambda_2, \ldots, \lambda_s$, and the DF that highly contributes to separating the means is Z_1, which corresponds to the first eigenvalue.
- Test the significance of each eigenvalue and keep only the significant eigenvalues. The test of significance requires the assumption of multivariate normality, which is not required for the DF. There are several tests, such as the Wilks and chi-square test, to determine the significance of each DF. Statistical packages provide the results of these tests. A significant result indicates that DF contributes significantly to separating the groups and forms one dimension.
- It is better to use a standardized DF if the variables are not commensurate.
- Interpreting the results means assessing the contribution of the variables; i.e., how the functions separate the groups. It is helpful to plot the scores of (the values of the DF for each of n observations) each DF and the first two DFs in a scatter diagram.

Note
- The first two or three DFs will be sufficient to capture almost all the differences among the groups.
- The relative importance of each DF is $\frac{\lambda_i}{\sum_{i=1}^{s} \lambda_i}$
- DA does not always produce useful results.

10.4 CLASSIFICATION

The second objective of DA is the prediction or allocation of observations to groups using discriminant variables to predict group membership. The rule for assigning individuals (observations) to one of the predefined groups with regard to a categorical variable (such as location, type of material) used to represent groups will be given in this section. Classification can be conducted either with linear classification functions or quadratic classification functions based on the covariance matrix of the m populations.

The procedure for classification or allocation of an observation to a group is based on calculating the Mahalanobis distance for each observation from a group center.

Suppose there are m groups (samples) and k variables measured from each sample. The procedure for assigning observations to a group is given below:

- Calculate the mean vector for each group ($\overline{Y}_1, \overline{Y}_2, \ldots, \overline{Y}_k$).
- Calculate the distance
  a. Linear classification functions: the groups have equal covariance matrices.

$$D_i^2 = (Y - \overline{Y}_i)'S_p^{-1}(Y - \overline{Y}_i)'$$

Where S_p is the pooled variance.
  b. Quadratic classification functions: the groups have unequal covariance matrices

$$D_i^2 = (Y - \overline{Y}_i)'S_i^{-1}(Y - \overline{Y}_i)'$$

Where S_i is the sample covariance matrix for ith group.
- Assign (allocate) Y to the group for which D_i^2 has the smallest value.

Note
- The proportion of misclassification is usually calculated to assess the ability of classification procedure to predict group membership, and the proportion of correct classification is also calculated. For example, if p_1 represents correct classification, then $p_2 = 1 - p_1$ represents misclassification. Any statistical package provides all proportions with the group membership in the form of a table.
- Prior probabilities could be used for classification; otherwise, all groups are considered to have equal probabilities.

10.5   DISCRIMINANT ANALYSIS IN R

DA can be conducted easily and efficiently by using the built-in functions provided by the R statistical software with simple commands to obtain a certain result or details. DA requires the MASS package to be installed and loaded. The package can be downloaded from the R library by typing `install.packages("mass")` and then pressing Enter to start downloading the required package. The last step is to call the package with the call `library (MASS)`. Alternatively, you can download from the main menu of R on the upper panel (**Packages**), and in RStudio the package can be downloaded from **Tools** on the upper panel of the RStudio interface. R offers a variety of built-in functions to execute DA including linear classifications, quadratic classifications, and related functions. The functions and structure with sufficient explanations of each function will be presented in this section.

1. The function `lda()` is used to perform linear DA. Let $Y_1, Y_2, \ldots, Y_k$ be the k responses measured on each sample from different groups (the groups are represented by a categorical variable). This function is used to produce the group means, Prior probabilities of groups, and the Coefficients of linear discriminants.

```
lda (Group ~ Y1 + Y2 + ... + Yk, data = data frame)
```

Although the structure of this function is similar to regression analysis, there is a difference in the meaning of the variables. For instance, whereas Group is a categorical variable (grouping variable) in DA, in regression analysis it represents the dependent variables. Furthermore, Y1,..., Yk are not independent variables as in regression analysis. The similarity is only in the structure of the call. By default, this call will use the frequencies in each group as the prior probabilities for the classification. These prior probabilities can be selected to change the default setting. Thus, the structure of a vector of equal prior probabilities for m groups is $(1/m, 1/m, \ldots, 1/m)$.

```
lda ( Group ~ Y1 + Y2 + ... + Yk, data = Data frame, prior = c(1/m,1/m,...,1/m))
```

Alternatively, the prior probabilities can be set as desired.

2. The function `qda()` is used to perform quadratic DA. The structure for this call is similar to the linear DA. In fact, the only syntactical difference is to call `qda()` instead of `lda()`.

```
qda(Group ~ Y1 + Y2 + ... + Yk, data = file name, prior = c(1/3, 1/3, 1/3))
```

3. The function `$class` is used to produce the coefficients of the DF.

```
Data frame $ class
```

4. The function `predict()` uses the output results of the function `lda()` to produce the classification of the groups, the posterior probabilities of each sample for each group, and the scores for each DF.

```
predict (data frame)
```

5. The function `$x` is used to extract the scores of each DF from the output of the function `predict()`.

```
Data frame $ x
```

6. The function `$class` is used to extract the classification of the groups from the output of the function `predict()`.

```
Data frame $ class
```

7. The function `$posterior` is used to extract posterior probabilities (the probability of assigning observations to groups) from the output of the function `predict()`.

```
Data frame $ posterior
```

8. The function `table()` is used to produce the classification table.

```
table (data frame $ grouping variable, Scores $ class)
```

9. The function `prop.table ()` is used to produce the percentages of the correct classifications and misclassifications.

```
prop.table (data frame,1 or 2)
```

The number in the function `prop.table()` indicates how to calculate the percentages of correct classifications and misclassifications: 1 means calculate the percentage based on the row, and 2 means use the column percentage.

10. The percentage of only correct classifications can be produced by using the function `diag (prop.table())`.

```
diag (prop.table (data frame,1 or 2))
```

11. The total percentage for the correct classification can be calculated by using the function `sum(diag (prop.table ()))`, which represents the overall predictive accuracy (the proportion of cases that lie along the diagonal).

```
sum (diag (prop.table (data frame,1 or 2)))
```

Example 10.1: Assessment of heavy metals in cockles—Samples of cockles (*Anadara granosa*) obtained from two rivers in the Penang State of Malaysia were analyzed for the content of arsenic (As) and heavy metals (Cr, Cd, Zn, Cu, Pb, and Hg) using a graphite flame atomic absorption spectrometer (GF-AAS) for Cr, Cd, Zn, Cu, Pb, and As and a cold vapor atomic absorption spectrometer (CV-AAS) for Hg. The two locations of interest were Kuala Juru (the Juru River) and Bukit Tambun (the Jejawi River), with 20 sampling points from each location. The data obtained from the two locations are given in Table 1.3.

The data for the arsenic and heavy metal concentrations in cockles had been stored as a `.CSV (EXAMPLE10_1)` file so it could be imported into an R data frame (R environment). Importing the stored data can be achieved by using the commands provided in R and RStudio (`read.csv("Example10_1")`) to call the stored data for the arsenic and heavy metal concentrations in cockles and to implement the required analysis. The second way to call the stored data into RStudio is by clicking on Environment (on the top right panel of the screen) and then choosing "Import Dataset" in the main menu of RStudio.

There are two groups (locations) where the cockle samples were collected and tested: the Juru River represents the first (location), and the Jejawi River represents the second group (location). Only one DF can be obtained that describes the differences in the arsenic and heavy metal data between the two rivers and explains 100% of the total variance. The

DA was performed using the function `lda()`, and the results of this function were saved in a new place called `DA` to be used for other purposes, such as to generate a graph. The results of using the `lda()` function are shown below.

```
>install.packages("mass")
>library(MASS)
>DA <- lda(Location ~ Cr + As + Cd + Zn + Cu + Pb + Hg, data = Example10_1)
>DA
Call:
lda(Location ~ Cr + As + Cd + Zn + Cu + Pb + Hg, data = Example9_1)

Prior probabilities of groups:
Jejawi  Juru
  0.5   0.5

Group means:
            Cr      As      Cd      Zn      Cu      Pb      Hg
Jejawi  0.1655  2.6910  0.8685  0.1965  0.1875  0.1230  1.3610
Juru    0.1670  2.6715  0.8870  0.2170  0.1850  0.1215  1.3315

Coefficients of linear discriminants:
          LD1
Cr   25.555345
As   -1.551576
Cd   26.931507
Zn   39.419984
Cu   -9.949634
Pb  -24.826411
Hg   -1.143475
```

The first two rows of the output display the call to install and load the package `MASS`. The third row shows the command used to conduct DA and save the results produced by the function `lda()` in a new place called `DA`. The result of the `prior probabilities of groups` includes the location, which reflects the proportion of each group within the data set. These results are presented after the DA. The average value of each variable (`Group means`) is displayed after the prior probabilities of each group. The last results produced by the function `lda()` are the coefficients associated with each parameter (variable) for the linear DF `Coefficients of linear discriminants:LD1`. The DF for the arsenic and heavy metal content in cockles can be expressed in terms of Eq. (10.2):

$$DF = 25.56Cr - 1.55As + 26.93Cd + 39.42Zn - 9.95Cu - 24.83Pb - 1.143Hg \qquad (10.2)$$

The relative contribution of each variable can be assessed based on the coefficients of the DF. Zn showed the highest contribution to separating the two groups (locations) and accounted for most of the expected variations between the two groups of cockles obtained from the two rivers; Cd had the second-highest contribution after Zn. The relative contributions to the arsenic and heavy metal contents can be arranged in order: $Zn > Cd > Cr > Pb > Cu > As > Hg$. The function `predict()` was used to assign the samples to the groups (locations) `$class`. Then, we computed the posterior probabilities (`$posterior`) of each sample to each group and the scores of the DF (`$x`).

```
> Scores <- predict(DA)
> Scores
$class
 [1] Juru     Juru     Jejawi   Juru     Juru     Juru     Juru     Juru     Juru
[10] Juru     Juru     Jejawi   Jejawi   Juru     Juru     Juru     Juru     Juru
[19] Juru     Juru     Jejawi   Jejawi   Juru     Jejawi   Jejawi   Jejawi   Juru
[28] Jejawi   Juru     Jejawi   Juru     Jejawi   Juru     Juru     Jejawi   Jejawi
[37] Jejawi   Jejawi   Jejawi   Jejawi
```

(Continued)

(Continued)

```
Levels: JejawiJuru
$posterior
            Jejawi          Juru
1       0.391289011   0.60871099
2       0.469526421   0.53047358
3       0.566200684   0.43379932
4       0.002927704   0.99707230
5       0.321389544   0.67861046
6       0.305893949   0.69410605
7       0.398040150   0.60195985
8       0.194120300   0.80587970
9       0.048791767   0.95120823
10      0.240995317   0.75900468
11      0.462181484   0.53781852
12      0.700735676   0.29926432
13      0.590981570   0.40901843
14      0.120712987   0.87928701
15      0.409575275   0.59042472
16      0.497239008   0.50276099
17      0.080766759   0.91923324
18      0.142355799   0.85764420
19      0.318323396   0.68167660
20      0.154741111   0.84525889
21      0.708924026   0.29107597
22      0.951850573   0.04814943
23      0.406708701   0.59329130
24      0.965633561   0.03436644
25      0.951673335   0.04832666
26      0.712127277   0.28787272
27      0.359921579   0.64007842
28      0.895013953   0.10498605
29      0.254690746   0.74530925
30      0.789731649   0.21026835
31      0.191019602   0.80898040
32      0.523525670   0.47647433
33      0.486406138   0.51359386
34      0.479162112   0.52083789
35      0.955499702   0.04450030
36      0.743045116   0.25695488
37      0.867476180   0.13252382
38      0.737844424   0.26215558
39      0.706766777   0.29323322
40      0.934729065   0.06527093

$x
          LD1
1    0.300451392
2    0.082980315
3   -0.181105952
4    3.964301004
5    0.508162074
```

(Continued)

(Continued)

6	0.557110980
7	0.281237556
8	0.967825801
9	2.019456667
10	0.780016317
11	0.103049942
12	-0.578471960
13	-0.250224536
14	1.350099288
15	0.248658579
16	0.007509009
17	1.653529791
18	1.221027510
19	0.517744883
20	1.154416285
21	-0.605233665
22	-2.028926084
23	0.256727004
24	-2.267986795
25	-2.026301315
26	-0.615822726
27	0.391428706
28	-1.457060541
29	0.730055495
30	-0.899734072
31	0.981384771
32	-0.064028866
33	0.036979660
34	0.056704606
35	-2.085113883
36	-0.721969330
37	-1.277437038
38	-0.703569961
39	-0.598141093
40	-1.809729816

The output of the function `predict()` starts with the function `$class`, which was used to classify each sample and assign the samples to the groups. The second result produced from the function `predict()` is a list of the posterior probabilities of each sample with respect to each group; this list measures the strength of the classification and can be extracted by the function `$posterior`. The samples will be assigned to the group with a higher probability than the others. The last results of the `predict()` function are the scores of each sample for the DF (`$x`).

It is necessary and helpful to plot the DF and study the relationship between the DF and the samples of cockles obtained from various locations. The function `barplot()` was used to produce a bar chart of the scores of the DF for different samples selected from both locations. A plot of the scores of the DF for various samples is given in Fig. 10.1 (see the appendix for the R code). The horizontal axes represent the cockle samples along 20 sampling points of each river, and the height represents the values (scores) of the DF for all samples. Fig. 10.1 reveals that most cockle samples obtained from the Juru River showed a positive contribution to the DF, and only three samples showed a negative contribution. In contrast, most of the cockle samples obtained from the Jejawi River showed a negative contribution, and only six samples showed a positive contribution. The positive contribution was mainly attributed to the high content of

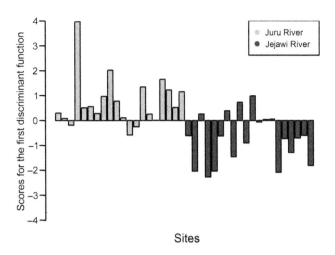

FIGURE 10.1   A bar chart for the scores of the discriminant function for all samples in both rivers.

Zn, Cd, and Cr in cockles, and the negative contribution was mainly attributed to the high content of Pb and Cu in cockles.

This pattern can be attributed to the differences in the industrial and agricultural activities operating close to the study areas. In general, the heavy metal contents such as those of Zn and Cd were higher in cockles obtained from the Juru River than from the Jejawi River.

Example 10.2: Classification of the heavy metal in cockles—Use the data for the arsenic and heavy metal content in cockles and the results of the DF obtained in Example 10.1 to illustrate the rule of classification. Then, construct a table for the numbers and percentages of correct classifications and misclassifications.

The classification procedure for all observations was carried out using (10.1) to assign all observations obtained from all locations to predefined groups (the Juru River and Jejawi River). Classification analysis can be carried out using R's built-in functions. The function `table()` was used to produce a table for the number of correct classifications and misclassifications for all of the observations. Similarly, the function `prop.table(data frame,1)` is used to produce the percentages of correct classifications and misclassifications for the classification operation with the data on the arsenic and heavy metal content in cockles. The table of the number of correct classifications and misclassifications and the table of the percentage values for the arsenic and heavy metal content in cockles obtained from different locations of the Juru and Jejawi Rivers were calculated using the functions `table()` and `prop.table(data frame, 1)`.

```
> cT < - table (Example10_1 $ Location, Scores $ class)
> cT
            Jejawi   Juru
   Jejawi      14       6
   Juru         3      17
> prop.table (cT, 1)
            Jejawi   Juru
   Jejawi     0.70   0.30
   Juru       0.15   0.85
> diag (prop.table (cT, 1))
   Jejawi  Juru
     0.70  0.85
> sum (diag(prop.table (cT)))
[1] 0.775
```

The results showed that 14 out of 20 of the samples from the Jejawi River were correctly classified into their group, which is equivalent to 70% of the Jejawi River samples, and $1 - 0.70 = 0.30$ represent misclassifications. Hence, the results of these samples were closer to the Juru River than the Jejawi River. However, 85% of the Juru River samples were correctly classified into the Juru River group (7 samples out of 20).

In summary, 77.5% of the samples collected from both rivers were correctly classified into their respective groups as given by the function `sum(diag(prop.table(cT, 1)))`. The results of the classification also showed that significant differences existed between these two rivers, which are expressed in terms of one DF.

Example 10.3: Quality characteristics for deep-fried snacks—The quality characteristics of oil extracted from the samples of tapioca deep-fried chips were studied in terms of the fat oxidation, fat content, and oil color. Seven different brands of tapioca chips were selected based on the packaging type (see Example 1.3 for more details on the packaging). The samples were analyzed for their fat oxidation (peroxide (PV), p-anisidine(PAV), acid value *(AV)* [mg KOH/g] and thiobarbituric acid (TBA) [mg MAD/kg]), fat content (%), fatty acid composition (%) (saturated fatty acid (SFA), poly-unsaturated fatty acid (PUFA), and monounsaturated fatty acid (MUFA)), oil color values (the lightness (L*) value and the redness (a*), yellowness (b*)), and moisture content (%). The data are given in Table 1.4.

In this example, there are seven different brands of tapioca chips whose firms use different packaging technology. The goal is to find the variables responsible for separating these brands based on the selected variables (i.e., to identify the variables responsible for the differences). The data had been stored as a `.CSV` (Example10_3) file to prepare it for analysis. DA was conducted for the data on tapioca chips, which include 12 parameters and 7 brands, using the function `lda()` in R.

```
> DA <- lda(Brand ~ FC + Mo. + L + a + b + PV + PAV + AV + TBA + SAF + MUFA + PUFA, data = Example10_3)
> DA
Call:
lda(Brand ~ FC + Mo. + L + a + b + PV + PAV + AV + TBA + SAF + MUFA + PUFA, data = Example10_3)

Prior probabilities of groups:
        1         2         3         4         5         6         7
0.1428571 0.1428571 0.1428571 0.1428571 0.1428571 0.1428571 0.1428571

Group means:
        FC       Mo.        L         a         b        PV        PAV        AV
1  20.18333  2.266667  73.00333  14.64333  83.56667  4.100000  13.876667  4.273333
2  18.58000  3.343333  23.76000  44.28333  40.73000 17.046667  10.063333 22.976667
3  17.47333  3.576667  79.43333  12.27000  48.82667  2.650000   6.736667  4.073333
4  26.23333  5.393333  70.04667  11.44000  77.16000 10.600000   8.530000  8.256667
5  20.67333  2.660000  82.34000  12.64333  75.19667  8.643333  11.966667  4.376667
6  30.01333  4.573333  67.25333  41.09000 114.56667 10.893333   2.543333  3.106667
7  25.87667  4.496667  72.84667  18.61333  99.23667  3.483333  14.806667  8.916667
        TBA       SAF      MUFA      PUFA
1  27.033333  50.25000  39.91000  9.843333
2  80.233333  49.48333  27.81667  8.513333
3   9.623333  51.55333  38.51667  9.933333
4  60.600000  50.93000  39.24333  9.830000
5  40.233333  52.25000  37.71000 10.073333
6   5.846667  51.60333  37.91333 10.483333
7   4.730000  49.03667  30.21000  9.243333

Coefficients of linear discriminants:
             LD1          LD2          LD3         LD4         LD5         LD6
FC     1.54355148     1.331383    1.3919115  -2.6204893  -3.4544387  -1.4229219
Mo.  -12.88296886    -9.268588  -17.5955563   0.2176488  -4.7148852   2.4560871
L   -205.72441901   -22.462598  -88.1176902   3.0647979  -2.0285901  -1.8554791
a    173.25819141    -1.159523 -112.6603983   3.0478668  -0.5840882  -1.9652839
b     27.75509249    99.425805   30.8419306  -0.2925046   0.7609744   0.4906181
PV    -1.00259116     0.496806    0.3226851   0.2792942  -0.8467373  -1.2252934
```

(Continued)

(Continued)

```
PAV   -9.27335214    -7.514609   -2.9995716    4.0487698    0.3682748   -3.6436545
AV     0.21728332    -2.137778    0.5248459   -0.3507663   -0.6257691   -0.7351909
TBA   -0.03641292     0.877998    0.4077783   -0.1095734   -0.2599982   -0.3451573
SAF    2.88292194    -9.474245    5.3643914   -6.9203379   -2.7873944   -1.3291017
MUFA  -6.48587210     3.615978   -4.3136722   -3.3612575    4.4603269   -0.5696028
PUFA   0.26806950   -64.380985   34.5341670  -18.7991045    2.1204217   -5.2965810

Proportion of trace:
   LD1     LD2     LD3     LD4     LD5     LD6
 0.8530  0.1344  0.0125  0.0000  0.0000  0.0000
```

Because the output of R for the data on tapioca chips is similar to the output of Example 10.1, there is no need to repeat the steps.

The DFs appeared under the `Coefficients of linear discriminants`. The result of the DA showed that the first three DFs explain 100% of the total variance (`Proportion of trace`) and are responsible for all of the differences among different brands of tapioca chips. In contrast, the other DFs explained 0% of the total variance. Therefore, only three DFs captured 100% of the differences among the 7 brands of tapioca deep-fried chips. The proportion of variance explained by each DF appeared under `LD1`, `LD2`, and `LD3` in the last row of the R output. The three DFs can be written in mathematical form as in (10.3−10.5).

$$DF_1 = 1.543FC - 12.882MO - 205.724L^* + 173.258a^* + 27.755b^* - 1.002PV - 9.273PAV + 0.217AV$$
$$- 0.036TBA + 2.882SAF - 6.485MUFA + 0.268PUFA \tag{10.3}$$

$$DF_2 = 1.331FC - 9.268MO - 22.462L^* - 1.159a^* + 99.425b^* - 0.496PV - 7.514PAV - 2.137AV$$
$$- 0.877TBA - 9.464SAF + 3.615MUFA - 64.380PUFA \tag{10.4}$$

$$DF_3 = 1.391FC - 17.595MO - 88.117L^* - 112.660a^* + 30.841b^* + 0.322PV - 2.999PAV + 0.524AV$$
$$+ 0.407TBA + 5.364SAF - 4.313MUFA + 34.534PUFA \tag{10.5}$$

The relative contributions of the selected variables in separating the seven brands of tapioca deep-fried chips can be assessed based on the coefficient associated with each variable in the DFs. The results of the first DF (`LD1`) showed that L* and a* exhibited the highest contribution to discriminating between different brands. In addition, b*, moisture, PAV, and MUFA accounted for most of the expected variations of the quality characteristics. However, the other parameters had much smaller contributions to explaining the variations between different brands. The first DF is responsible for 85.3% of the differences between the brands, and the second DF (`LD2`) is responsible for 13.44% of these differences. It is clear that b* showed the highest contribution, and it was followed by PUFA; the other parameters showed less contribution. The third DF (`LD3`) explained very little of the total variance (1.25%). Furthermore, the first two DFs are responsible for separating the brands and explained 98.74% of the total variation between them.

The results of using the function `predict()` for different brands of tapioca deep-fried chips including the classification of each sample (`$class`), posterior probabilities (`$posterior`), and the last results are the scores (`$x`) of each DF. These scores are presented below.

```
> Scores <- predict(DA)
> Scores is      # scores for each sample
$class
 [1] 1 1 1 2 2 2 3 3 3 4 4 4 5 5 5 6 6 6 7 7 7
Levels: 1 2 3 4 5 6 7

$posterior
     1  2  3  4  5  6  7
1    1  0  0  0  0  0  0
2    1  0  0  0  0  0  0
```

(Continued)

```
3    1  0  0  0  0  0  0
4    0  1  0  0  0  0  0
5    0  1  0  0  0  0  0
6    0  1  0  0  0  0  0
7    0  0  1  0  0  0  0
8    0  0  1  0  0  0  0
9    0  0  1  0  0  0  0
10   0  0  0  1  0  0  0
11   0  0  0  1  0  0  0
12   0  0  0  1  0  0  0
13   0  0  0  0  1  0  0
14   0  0  0  0  1  0  0
15   0  0  0  0  1  0  0
16   0  0  0  0  0  1  0
17   0  0  0  0  0  1  0
18   0  0  0  0  0  1  0
19   0  0  0  0  0  0  1
20   0  0  0  0  0  0  1
21   0  0  0  0  0  0  1
```

```
$x
            LD1           LD2          LD3          LD4
1     -2410.256     513.34081     502.1448    4.3721703
2     -2409.365     511.98173     504.0880    5.1438903
3     -2406.096     512.01128     502.2421    4.3705970
4     11752.929   -2596.07672     210.3537    8.3270879
5     11751.912   -2596.60902     209.5355    9.3124225
6     11752.520   -2595.31339     209.1734    5.6302751
7     -5046.067   -3084.72088    -864.0243    0.7031633
8     -5047.421   -3083.92362    -865.1968    1.8012230
9     -5046.843   -3083.79178    -865.2010    1.3558536
10    -2514.643     -19.04139     921.4623  -54.6722836
11    -2515.579     -18.17615     919.9198  -55.3743370
12    -2515.044     -19.69528     918.7281  -53.8046115
13    -4878.866    -543.05077    -318.4334    9.5060109
14    -4880.698    -543.85045    -317.0763    9.1902919
15    -4879.522    -544.74455    -319.0055   10.3480162
16     4317.741    3695.46937    -977.4038  -23.0395391
17     4317.820    3694.48817    -975.2192  -23.5255949
18     4318.850    3698.14493    -975.0319  -21.4041656
19    -1220.359    2034.85076     526.4367   53.0462364
20    -1220.432    2035.22371     525.6500   54.6365924
21    -1220.580    2033.48324     526.8578   54.0767012
             LD5           LD6
1      41.66211112   -1.6747892
2      41.33617085   -0.3962827
3      40.41199784   -0.7359158
4      -1.16187538   -4.0746627
5      -1.83620533   -2.3710680
6       0.17071935   -2.6664619
7       0.12483936   14.2387858
```

(Continued)

(Continued)

8	-0.60278524	15.1745219
9	-0.05857082	13.6687518
10	-13.55213169	2.6001985
11	-17.60033867	2.4562022
12	-14.15019882	4.9112999
13	-9.65485843	-23.7067470
14	-9.45636838	-23.2883254
15	-9.38502353	-25.0937346
16	1.54684831	0.8475652
17	0.48659844	2.4192273
18	1.27552459	1.9397224
19	-17.07772011	9.9040454
20	-15.99201082	8.5622044
21	-16.48672264	7.2854624

The bar chart in Fig. 10.2 represents the scores of the first DF for different samples obtained from different brands of tapioca chips; it was produced using the function `barplot()` in R. Observe in Fig. 10.2 that the behavior of the seven brands is different with regard to the selected parameters (variables), as the samples fluctuated between a positive and negative contribution to the first DF. The highest positive contribution was exhibited by brand 2, and brand 6 made the second greatest contribution. Both the third and fifth brands exhibited the highest negative contribution, and brands 1, 3, and 7 exhibited negative contributions that were smaller than the third and fifth brands. The positive contribution is mainly due to a* and b*, whereas the negative contribution is mainly attributed to L*; much less of the contribution was due to the other parameters. Different behavior for the selected parameters regarding the second DF was exhibited in Fig. 10.3; the positive contribution is mainly due to b*. Furthermore, some brands exhibited a negative contribution, which is mainly due to the high values of PUFA and L* and less to the other parameters. This difference in the behavior of different types of tapioca deep-fried chips indicates that different brands have different properties.

It is useful to place the scores of the first two DFs in a scatter diagram to obtain a clear picture of the behavior of different brands of tapioca chips with respect to the first two DFs. The function `plot()` is used to produce a plot of the scores of the first two DFs, as in Fig. 10.4 (see the appendix for the R code). The difference is very clear in the behavior of different brands of tapioca chips with regard to the first two DFs.

Example 10.4: Classification of quality characteristics for deep-fried snacks—Use the data on the arsenic and heavy metal content in cockles and the results of the DF obtained in Example 10.3 to illustrate the rule of classification. Construct a table of the numbers and percentages of the correct classifications and misclassifications.

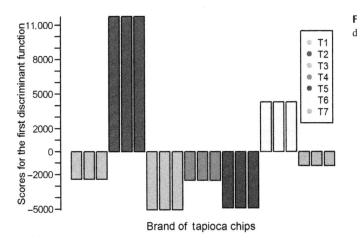

FIGURE 10.2 Scores for the first discriminant function for tapioca deep-fried chips.

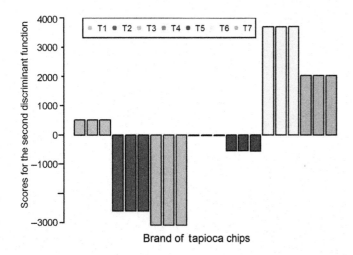

FIGURE 10.3 Scores for the second discriminant function for tapioca deep-fried chips.

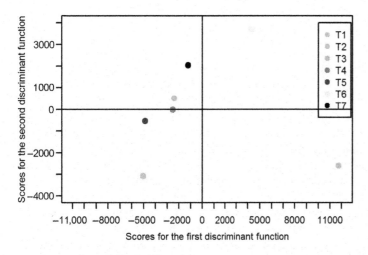

FIGURE 10.4 Scores for the first and second discriminant functions for tapioca chips.

The rule of classification was carried out on the data given in Example 10.3, which included seven brands of tapioca deep-fried chips and 12 parameters, using the functions `table()`, `prob.table(data frame, 1)`, and `sum(diag(prop. table(data frame, 1)))`.

```
> cT <- table (Example10_3 $ Brand, Scores $ class)
> cT
    1 2 3 4 5 6 7
  1 3 0 0 0 0 0 0
  2 0 3 0 0 0 0 0
  3 0 0 3 0 0 0 0
  4 0 0 0 3 0 0 0
  5 0 0 0 0 3 0 0
  6 0 0 0 0 0 3 0
  7 0 0 0 0 0 0 3
> prop.table (cT, 1)
    1 2 3 4 5 6 7
  1 1 0 0 0 0 0 0
  2 0 1 0 0 0 0 0
  3 0 0 1 0 0 0 0
  4 0 0 0 1 0 0 0
```

(Continued)

```
(Continued)
   5  0  0  0  0  1  0  0
   6  0  0  0  0  0  1  0
   7  0  0  0  0  0  0  1
>diag(prop.table(cT,1))
1 2 3 4 5 6 7
1 1 1 1 1 1 1
>sum(diag(prop.table(cT)))
[1] 1
```

There was a 100% correct classification rate for the seven brands of tapioca deep-fried chips and a 0% misclassification rate. This strong result indicates that each brand of tapioca chip is affected by different conditions. The difference could be due to different ingredients, the technology used for packaging, or other reasons.

FURTHER READING

Alkarkhi, A. F. M., Ismail, N., & Easa, A. M. (2008). Assessment of arsenic and heavy metal contents in cockles (*Anadara granosa*) using multivariate statistical techniques. *Journal of Hazardous Materials, 150*, 783–789.

Alkarkhi, A. M., Alqaraquli, W. A. A., Nurl, H., & Yusri, Y. Tapioca. Unpublished work.

Bryan, F. J. M. (1991). *Multivariate statistical methods: A primer*. Great Britain: Chapman & Hall.

Chi, Y. (2017). *R Tutorial, An introduction to R, Hierarchical Cluster Analysis* [Online]. Available: <http://www.r-tutor.com/gpu-computing/clustering/hierarchical-cluster-analysis> Accessed 27.08.17.

Daniel, H. (2013). *Blog Archives, High Resolution Figures in R* [Online]. Available: <https://www.r-bloggers.com/author/daniel-hocking/> Accessed 15.07.17.

Johnson, R. A., & Wichern, D. W. (2002). *Applied multivariate statistical analysis*. New Jersey: Prentice Hall.

Rencher, A. C. (2002). *Methods of multivariate analysis*. New York: J. Wiley.

Robert, I. K. (2017). *Quick-R, Accessing the Power of R, Discriminant Function Analysis* [Online]. Available: <http://www.statmethods.net/advstats/discriminant.html> Accessed 26.08.17.

Technical University of Denmark (DTU). (2015). *Discriminant Analysis: LDA, QDA, k-NN, Bayes, PLSDA, cart, Random forests* [Online]. DENMARK. Available: <http://27411.compute.dtu.dk/enote/afsnit/NUID202/> Accessed 27.08.17.

Thiagogm. (2014). *Computing and Visualizing LDA in R* [Online]. R-bloggers. Available: <https://www.r-bloggers.com/computing-and-visualizing-lda-in-r/> Accessed 27.08.17.

Chapter 11

Cluster Analysis

Learning outcomes

At the end of this chapter, you should be able

- *To describe cluster analysis (CA).*
- *To summarize the results in a plot called a dendrogram.*
- *To know how to apply CA to food science data.*
- *To apply R's commands and built-in functions in performing CA.*
- *To understand and interpret the R output for CA.*
- *To report on the useful conclusions.*

11.1 INTRODUCTION

Cluster analysis (CA) is a multivariate tool used to organize a set of multivariate data (observations, objects) into groups called clusters. The observations within each group are close to each other (similar observations); however, the clusters themselves are dissimilar. There are a number of algorithms for sorting data into groups based on similarity, and they are significantly different. Clustering techniques have been applied to a wide variety of research problems where there is a need to sort a huge data set into groups. The number of groups and the number of observations in each cluster are unknown before starting the grouping process. Cluster analysis is valuable for classifying and identifying the true groups. For instance, a large number of deep-fried-chip samples are obtained from different brands, and the researcher wants to know how many different brands can be similar. If the brands can be grouped in clusters, then a sample could be selected from each cluster to study the activities around the new groups as long as the brands are similar within the same cluster and differ with other clusters.

Cluster analysis is different from classification analysis because in classification analysis, the number of groups is known and the objective is to reassign each observation to one predefined group, while in cluster analysis, there is no assumption of predefined groups and the number of groups will be determined based on the similarity between the observations (items). Cluster analysis has been used in many fields such as medicine, anthropology, geology, environment, food science, and engineering.

11.2 MEASURES OF DISTANCE

Cluster analysis is usually used to group multivariate data on k variables $X_1, X_2, \ldots, X_k$ using an index of proximity, which is generally indicated by a distance. Cluster analysis groups similar observations in one group, which means observations that are close to each other form a group. The distance is a measure of the dissimilarity.

The Euclidean distance function is usually used to measure the distance between the individuals. The distance between two points A and B with coordinates $A = (X_1, X_2, X_k)$ and $B = (Y_1, Y_2, \ldots, Y_k)$ is called the Euclidean distance, as defined in (11.1).

$$d = \sqrt{(X_1 - Y_1)^2 + (X_2 - Y_2)^2 + \ldots + (X_k - Y_k)^2} \tag{11.1}$$

Variables are usually standardized to take into account the variation and the correlation (or covariances) between different variables. The formula for the statistical distance as given in (11.2) is

Easy Statistics for Food Science with R. DOI: https://doi.org/10.1016/B978-0-12-814262-2.00011-X

$$d = \sqrt{\frac{(X_1 - Y_1)^2}{S_{11}} + \frac{(X_2 - Y_2)^2}{S_{22}} + \ldots + \frac{(X_k - Y_k)^2}{S_{kk}}} \qquad (11.2)$$

11.3   APPROACHES FOR CLUSTERING

Many algorithms have been proposed for clustering. All proposed algorithms provide clusters such that the resulting groups will reflect the data structure. The approaches for clustering can be classified into two types:

11.3.1   Hierarchical Clustering

The methods used for hierarchical clustering can be classified into two types: agglomerative and divisive approaches. Hierarchical clustering is a sequential process. These methods start by calculating the distances of each individual to all the other individuals and forming a matrix called the distance matrix for all individuals. The result of hierarchical methods is usually summarized in a figure called a dendrogram, as given in Fig. 11.1.

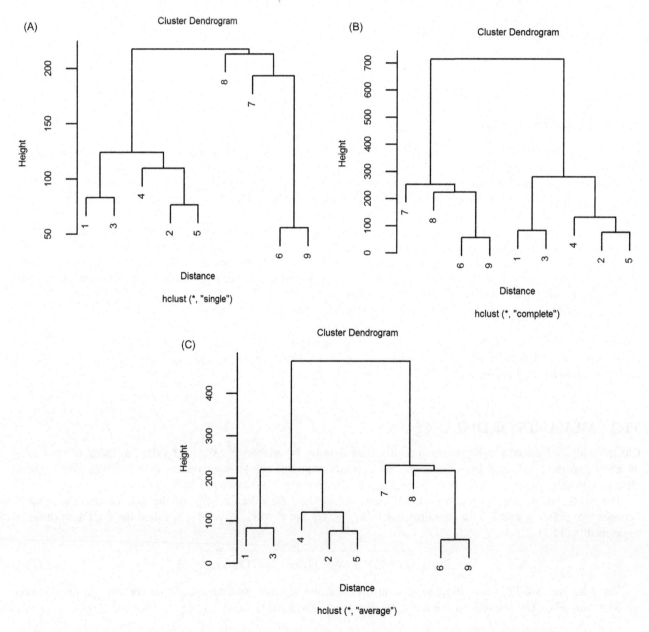

FIGURE 11.1 Dendrogram showing hierarchical analysis using (A) single linkage; (B) complete linkage; and (C) average linkage.

11.3.1.1 Agglomerative Approach

The agglomerative approach starts with n clusters, with one individual in each cluster, and then groups that are close to each other are merged to form a new cluster. The number of clusters becomes lower, and the number of observations in each cluster becomes higher or the cluster becomes larger. This approach ends with one cluster containing all the individuals. At each step, the distance is calculated between all the pairs.

11.3.1.2 Divisive Approach

This approach works in a direction opposite to that of the agglomerative approach. In this approach, all individuals start with a single cluster containing all individuals. The algorithm then partitions the cluster into two clusters; at each step, one of the two clusters (groups) is partitioned into two subgroups. The process continues until stability is achieved.

In general, the agglomerative approach is more common than the divisive approach. The hierarchical methods are single linkage (nearest neighbor), complete linkage (farthest neighbor), average linkage, median, ward's method, and the flexible data method.

11.3.2 Nonhierarchical Methods

Nonhierarchical methods use some criteria other than distance to cluster the observations into a certain number of clusters, say g. The methods for nonhierarchical clustering are partitioning, k-means, and mixture methods.

11.4 CLUSTER ANALYSIS IN R

R statistical software provides built-in functions to conduct cluster analysis and produce beautiful graphs. The functions used to perform cluster analysis with the structure for each function are presented in an easy and simple way.

1. The function `dist()` is used to calculate the Euclidean distance matrix between any two values in a dataset. The data for the distance matrix should be stored in a new place to be used for further analysis.

```
dist (data frame)
```

2. The function `hclust()` is used to perform hierarchical analysis on a distance matrix calculated by the function `dist()`. The results should be saved in a new place for further use.

```
hclust (Data frame)
```

Several methods are available in R for clustering, including `"single"`, `"average"`, `"complete"`, `"ward"`, and others. R usually uses the complete method by default. The structure of the call for clustering with a specification of the method of clustering is given below.

```
hclust (Data frame, method = "complete")
```

3. The results of using the two functions `dist()` and `hclust()` can best be presented in a graph called a dendrogram. The function `plot()` in R can be used to produce the dendrogram and visualize how clusters are formed.

```
plot (Data frame)
```

A label can be added to the plot command.

```
plot (Data frame, labels = as.character (file name $ variable name))
```

4. The function `rect.hclust()` is used to identify a given number of clusters. This function is usually used after the `plot()` function to produce a dendrogram.

```
plot(data frame)    rect.hclust (Data frame, k)
```

where `k` is the number of clusters to be identified.
Alternatively, the dendrogram can be cut off at a specific height by adding the `h` argument to represent the height.

```
plot (Data frame)    rect.hclust (Data frame, h = height)
```

5. The function `cutree()` is used to save the cluster numbers to a new variable in the data set.

```
Dataset $ cluster number <- cutree (Data frame, n)
```

6. The function `kmeans()` is used to execute non-hierarchical clustering. The number of clusters should be identified.

```
kmeans(Data frame, number of clusters needed)
```

7. The function `aggregate()` is used to calculate the cluster means.

```
aggregate(Data frame, by = list( $cluster),FUN = mean)
```

Example 11.1: Flour prepared from green and ripe banana fruits - Flour prepared from green and ripe Cavendish (CGP and CRP) and Dream (DGP and DRP) banana fruits were assessed for starch and dietary fiber components such as the total starch (*TS*), digestible starch (*DS*), resistant starch (*RS*), total dietary fiber (*TDF*), soluble dietary fiber (*SDF*), and insoluble dietary fiber (*IDF*). The data for the starch and dietary fiber components of banana flour (%) prepared from fruits of different varieties and stages of ripeness are given in Table 11.1.

TABLE 11.1 Starch and Dietary Fiber Components of Banana Flour (%) Prepared From Fruits of Different Varieties and Stages of Ripeness

Variety	IDF	SDF	TDF	RS	DS	TS
CGP	11.91	3.58	15.50	52.88	20.61	73.49
CRP	15.25	3.75	18.99	12.19	6.19	18.39
DGP	10.91	3.33	14.25	50.62	18.43	69.05
DRP	14.27	3.47	17.74	11.74	5.71	17.45

The first step in clustering analysis is to calculate the distances between the different observations. The Euclidean distance formula is used to produce the distance matrix between the different varieties of flour prepared from banana fruits.

The data for the types of flour prepared from banana fruits had been stored as a .CSV (Example11_1) file. The Euclidean distances between various varieties of banana flour should be calculated first to prepare the data for cluster analysis. The distance matrix for all possible combinations between different varieties of banana flour was calculated using the function dist() provided in R.

```
> Distance <- round (dist (Example11_1[2:7]), digits = 2)
> Distance
           1      2      3
2    70.16
3     5.67  65.07
4    71.17   1.98  66.03
```

The R outputs showed a call of dist() to calculate the Euclidean distances between different varieties based on the selected parameters in the fileExample11_1[2:7]. The results were rounded to two decimals using the function round() and saved in a new place called Distance. The following numbers in the output represent the type of flour: 1 = CGP, 2 = CRP, 3 = DGP, and 4 = DRP. The next step is to perform cluster analysis for the flour prepared from green and ripe banana fruits by calling the distance matrix (which we calculated earlier). This matrix will be used by the function hclust() to conduct cluster analysis.

```
> CA <- hclust (Distance, method = "single")
> CA
Call:
hclust(d = Distance, method = "single")
Cluster method    : single
Distance          : euclidean
Number of objects : 4
```

The first row of the R output from using the function hclust() showed a call to execute cluster analysis using the single linkage method and save the results produced by this function in a new place called CA. The single linkage (nearest neighbor) method was used to illustrate the use of distances in the process of cluster analysis and show how the agglomeration occurs. Researchers can select any method.

The distance matrix has been placed in a table form (Table 11.2), and the smallest distance is **1.98** (bolded, Table 11.2); it corresponds to the distance between the varieties 2 = CRP and 4 = DRP (the highlighted column and row). Thus, CRP and DRP will form the first cluster: C1 = [CRP, DRP].

A new table of distances for different varieties of flour prepared from banana fruits is given in Table 11.3. This table has one fewer column and one fewer row than Table 11.2.

TABLE 11.2 Distance Matrix for Different Types of Flour

	CGP	CRP	DGP	DRP
CGP	0.00	70.16	5.67	71.17
CRP	70.16	0.00	65.07	**1.98**
DGP	5.67	65.07	0.00	66.03
DRP	71.17	**1.98**	66.03	0.00

TABLE 11.3 Shows the Distances After Forming the First Cluster

	CGP	C1	DGP
CGP	0.00	70.16	**5.67**
C1	70.16	0.00	65.07
DGP	**5.67**	65.07	0.00

TABLE 11.4 Shows the Distances After Forming the First and Second Clusters

	C1	C2
C1	0.00	**65.07**
C2	**65.07**	0.00

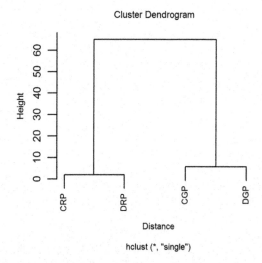

FIGURE 11.2   Dendrogram for the single linkage method for different types of flour.

The second step is to find the smallest distance in the new table (Table 11.3) to form the second cluster; it can be observed that the smallest distance is **5.67** (bolded) between 1 = CGP and 3 = DGP (the highlighted column and row). These points will form the second cluster C2 = [CGP, DGP]. The new table of distances is given in Table 11.4, and as before, it has one fewer column and one fewer row from the table that preceded it, namely, Table 11.3.

The last cluster C3 is formed from the first and second clusters: C3 = [C1, C2] (the highlighted column and row) because the smallest distance is **65.07**. The last cluster is cluster 3, which consists of all varieties and ends with one cluster for all varieties.

The steps of clustering analysis for various varieties of flour prepared from banana fruits can be summarized in a graph called a dendrogram. The function `plot()` was used to produce a dendrogram for the flour prepared from green and ripe banana fruits; see Fig. 11.2.

```
>plot(CA, labels = Example11_1 $ Variety,font.main = 1,
cex.lab = 0.5, cex.main = .5,font.sub = 1, cex.sub = .5,
cex.axis = 0.5, cex = 0.5)
```

The steps for the single linkage (nearest neighbor) method are presented in Fig. 11.2. The first cluster was formed by observation 2 (CRP) and observation 4 (DRP), and the second cluster was formed by observation 1 (CGP) and observation 3 (DGP). Then, the two clusters were merged to form the last cluster C3.

In summary, we determined that green fruits are close to each other and ripe fruits are close to each other. However, green fruits are different from ripe fruits, as can easily be observed in Fig. 11.2.

Therefore, only two groups or clusters were identified; the first cluster is of green banana fruits, and the second cluster is of ripe banana fruit regardless of the variety of banana. This result indicates that fruits of the same type have the same conditions regardless of the variety.

Note:

- The classification can be carried out for the variables or the observations.
- Researchers are free to decide the number of clusters to be considered. Knowledge and experience about the nature of the data should be considered to choose the number of clusters.

Example 11.2: Quality characteristics of deep-fried food - A researcher wants to investigate the quality characteristics of deep-fried snack products in the Malaysian market. Five varieties of deep-fried snack (muruku chips, tapioca chips, prawn crackers, fish crackers, and banana chips) were selected and analyzed for the food quality characteristics of the extracted oil: the fat content, moisture content, oil color values, fatty acid composition, and fat oxidation parameter. Seven different brands were selected based on the packaging technology. The samples were analyzed for their fat oxidation (Peroxide (PV), p-anisidine (p-AV), acid value (AV) [mg KOH/g] and thiobarbituric acid (TBA) [mg MAD/kg]), fat content (%) (FC), fatty acid composition (%) (saturated fatty acid (SFA), polyunsaturated fatty acid (PUFA), and monounsaturated fatty acid (MUFA)), oil color values (lightness (L*) value, redness (a*) and yellowness (b*)), and moisture content (%) (Mo). The data are given in Table 11.5.

The data for the assessment of the quality characteristics of deep-fried food with 12 parameters measured from each sample were stored as a `.CSV` (`Example11_2`) file. As usual in hierarchical clustering, the distances between different types of deep-fried snacks were calculated using the function `dist()`. The outputs of this function were placed in a matrix form and saved in a new place called `Distance` to be ready for the next step in the analysis. The two functions `dist()` and `hcclust()` were used to calculate the distances, and we conducted cluster analysis for different types of selected deep-fried snacks as shown below.

TABLE 11.5 Quality Characteristics of Oil Extracted From Deep-Fried Samples

Type	FC	Mo	L*	a*	b*	PV	p-AV	AV	TBA	SFA	MUFA	PUFA
Muruku	35.32	3.24	65.38	27.58	94.29	6.30	7.68	6.74	0.03	44.03	37.88	10.70
Tapioca	22.72	3.76	66.95	22.14	77.04	8.20	9.79	8.00	32.61	50.73	35.90	9.70
Prawn	19.56	4.50	79.65	16.58	95.77	3.75	5.88	4.37	14.15	48.73	42.55	10.73
Fish	31.64	3.68	79.80	1.52	62.50	1.27	16.16	3.41	6.63	45.70	43.07	11.29
Banana	36.90	3.05	76.74	16.54	122.03	2.98	10.71	3.15	5.10	44.84	42.20	12.67

```
> Distance <- round (dist (Example11_2[2:13]),digits = 2)
> Distance
        1      2       3      4
2   40.12
3   28.90  31.49
4   45.75  41.78   40.74
5   33.24  56.96   33.57  62.01
> CA <- hclust (Distance,"single")
> CA

Call:
hclust(d = Distance, method = "single")

Cluster method    : single
Distance          : euclidean
Number of objects: 5
```

The numbers in the outputs of the `Distance` matrix refer to the type of product, i.e., 1 = muruku, 2 = tapioca, 3 = prawn, 4 = fish, and 5 = banana. The results of all the possible distances are placed in table form (Table 10.4) for the analysis.

Because the single linkage (nearest neighbor) method was used to conduct the cluster analysis, the smallest distance in Table 11.6 should be chosen to form the first cluster. The smallest distance in the distance matrix (Table 11.6) is **28.90** (bolded), which represents the distance between muruku chips and prawn crackers (the highlighted column and row in Table 11.6). The first cluster (C1) is formed from muruku chips and prawn crackers, and the new table of distances will have fewer columns and one fewer row inasmuch as muruku chips and prawn crackers have been combined to represent cluster 1 (C1) in the new table, as shown in Table 11.7.

TABLE 11.6 Distance Matrix for Different Varieties of Deep-Fried Snacks

	Muruku	Tapioca	Prawn	Fish	Banana
Muruku	0.00	40.12	**28.90**	45.75	33.24
Tapioca	40.12	0.00	31.49	41.78	56.96
Prawn	**28.90**	31.49	0.00	40.74	33.57
Fish	45.75	41.78	40.74	0.00	62.01
Banana	33.24	56.96	33.57	62.01	0.00

TABLE 11.7 The Distances After Forming the First Cluster

	Tapioca	C1	Fish	Banana
Tapioca	0.00	**31.49**	41.78	56.96
C1	**31.49**	0.00	40.74	33.24
Fish	41.78	40.74	0.00	62.01
Banana	56.96	33.24	62.01	0.00

TABLE 11.8 The Distances After Forming the First and Second Clusters

	C2	Fish	Banana
C2	0.00	40.74	**33.24**
Fish	40.74	0.00	62.01
Banana	**33.24**	62.01	0.00

TABLE 11.9 The Distances After Forming the First Three Clusters

C3	0.00	**40.74**
Fish	**40.74**	0.00

TABLE 11.10 The Smallest Distances for all Steps of the Clustering

Stage	Smallest Distance
1	28.90
2	31.49
3	33.24
4	40.74

The second step is to search for the smallest distance in Table 11.7, which is **31.49** (bolded) between tapioca chips and C1 (the highlighted column and row in Table 11.7). These categories are combined to form the second cluster (C2). The new table of distances will have three rows and three columns because C1 and tapioca chips were merged to form the second cluster (C2), as shown in Table 11.8.

The next step is to seek the smallest distance in the new table (Table 11.8). The smallest distance is **33.24** (bolded) between cluster 2 (C2) and banana chips (the highlighted column and row), which are combined to form the third cluster (C3). Only two columns and two rows are left in the distance matrix in Table 11.9.

The last cluster is formed between cluster 3 (C3) and fish crackers, as the smallest distance is **40.74**. The smallest distances for all clusters are given in Table 11.10.

The results of the previous steps of the cluster analysis are summarized in a graph to illustrate the steps of the clustering and to convey the similarities among various types of deep-fried food. The function plot() was used to produce a dendrogram, as shown in Fig. 11.3.

```
>.plot(hc, labels = Example11_2 $ Brand,font.main = 1, cex.lab = 0.5, cex.main = .5,font.sub = 1,
cex.sub = .5, cex.axis = 0.5, cex = 0.5)
```

In summary, it can be concluded that all of the selected varieties of deep-fried food are different and their dissimilarity is very high. The differences in the quality parameters between deep-fried products could be attributed to the interplay between the processing practices, storage conditions, raw materials (including frying oils), packaging materials, and technologies.

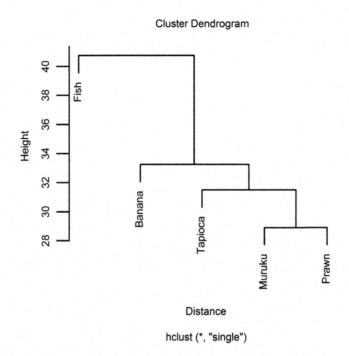

FIGURE 11.3   Dendrogram for the single linkage method for different types of selected snacks.

FURTHER READING

Alkarkhi, A. F. M., Muhammad, N. A. N., Alqaraghuli, W. A. A., Yusup, Y., Easa, A. M., & Huda, N. (2017). An investigation of food quality and oil stability indices of muruku by cluster analysis and discriminant analysis. *International Journal on Advanced Science, Engineering and Information Technology, 7*, 2279–2285.

Alkarkhi, A. F. M., Alqaraghuli, W. A. A., & Yusri, Y. Deep fried snacks.

Bryan, F. J. M. (1991). *Multivariate statistical methods: A primer.* Great Britain: Chapman&Hall.

Chi, Y. (2017). *R Tutorial, An R Introduction to statistics, Hierarchical Cluster Analysis* [Online]. Available: <http://www.r-tutor.com/gpu-computing/clustering/hierarchical-cluster-analysis> Accessed 27.08.17.

Daniel, H. (2013). *Blog Archives, High Resolution Figures in R* [Online]. Available: <https://www.r-bloggers.com/author/daniel-hocking/> Accessed 15.07.17.

Johnson, R. A., & Wichern, D. W. (2002). *Applied multivariate statistical analysis.* New Jersey: Prentice Hall.

Rencher, A. C. (2002). *Methods of multivariate analysis.* New York: J. Wiley.

Robert, I. K. (2017). *QuickR, accessing the power of R, Cluster Analysis* [Online]. Available: <http://www.statmethods.net/advstats/cluster.html> Accessed 27.08.17.

Saifullah, R., Noryati, I., Alkarkhi, A. F. M., & Azhar, M. E. (2010). The use of principal component and cluster analysis to differentiate banana peel flours based on their starch and dietary fibre components. *Tropical Life Sciences Research, 21*, 91–100.

Teja, K. (2017). *K Means Clustering in R* [Online]. R-Bloggers. Accessed 15.08.17.

Appendix

Chapter 1

Normal curve

```
x = seq (-3,3,0.1)
y = dnorm (x)
jpeg ("Figure 1.1.jpeg", res = 400,height = 8, width = 8, units = 'cm')
par (mar = c(5, 1, 3, 1))
plot (x, y, xlab = "", type = "l", lwd = 2, col = "black", xaxt = "n",yaxt = "n", ylab = "") # Make plot
x = seq (0, 0, length = 200)
y = dnorm (x)
segments (x, rep (0,length (x)), x, dnorm (x, 0, 1), lwd = 2)
mtext (c("μ"), side = 1, line = 1, at = c (0,0))
dev.off()
```

Chapter 2

Example 2.17

```
pH <- c(7, 4, 3, 9, 2)
jpeg ("Figure 2.9.jpeg", res = 400, height = 8, width = 8, units = 'cm')
par (mar = c(4.5, 2, 3, 1))
plot (pH, ylab = "", xlab = "", pch = 19, cex = 0.5, xaxt = "n", yaxt = "n")
axis (1, seq(1:5), cex.axis = 0.5, tck = -0.03, mgp = c (3, .3, 0))
axis (2, seq(1:10), cex.axis = 0.5, tck = -0.03, mgp = c (3, .3, 0))
mtext (side = 2, "pH", line = 1, cex = 0.5 )
mtext (side = 1, "index", line = 1, cex = 0.5 )
dev.off()
```

Example 2.18

```
pH <- c(7, 4, 3, 9, 2)
jpeg ("Figure 2.10.jpeg", res = 400, height = 8, width = 8, units = 'cm')
par (mar = c (4.5, 2, 3, 1))
plot (pH, type = "o", col = "black", ylab = "", xlab = "", pch = 19, cex = 0.5, xaxt = "n", yaxt = "n")
axis (1, seq(1:5), cex.axis = 0.5, tck = -0.03, mgp = c (3, .3, 0))
axis (2, seq(1:10), cex.axis = 0.5, tck = -0.03, mgp = c(3, .3, 0 ))
mtext (side = 2, "pH", line = 1, cex = 0.5 )
mtext (side = 1, "index", line = 1, cex = 0.5 )
dev.off()
```

Chapter 4

Example 4.1

```
# a
mean (Example4_1 $ pH)

# b
# key in the data
pH < - c(4.64, 4.32, 4.60, 4.59, 4.55, 4.33, 4.30, 5.08, 5.31, 5.33, 5.24, 5.26)
mean (pH)

#c
#more details
pH < - c(4.64, 4.32, 4.60, 4.59, 4.55, 4.33, 4.30, 5.08, 5.31, 5.33, 5.24, 5.26)
s = sum (pH)
s                    #print s
n = length (pH)
n                    #print n
Mean = s / n
Mean               #print Mean
```

Example 4.2

```
AV <- c(4.38, 4.48, 3.96, 24.41, 22.32, 22.2, 4.05, 3.96, 4.21, 7.97, 8.24, 8.56, 4.47, 4.2, 4.46, 3.42,
2.99, 2.91, 9.93, 8.51, 8.31)
mean (AV)

#more details
AV <- c(4.38, 4.48, 3.96, 24.41, 22.32, 22.2, 4.05, 3.96, 4.21, 7.97, 8.24, 8.56, 4.47, 4.2, 4.46, 3.42,
2.99, 2.91, 9.93, 8.51, 8.31)
s = sum (AV)
s                    #print s
n = length (AV)
n                    #print n
Mean = s/n
Mean               #print Mean
```

Example 4.3

```
Mean <- sapply (Example4_3, mean)
Mean
```

Example 4.4

```
Mean <- sapply (Example4_4, mean)
Mean
```

Example 4.5

```
Var (Example4_5 $ pH)
Sd (Example4_5 $ pH)
```

Example 4.6

```
Var (Example4_6 $ AV)
Sd (Example4_6 $ AV)
```

Example 4.7

```
round (cov (Example4_7), digits = 2)
round (cor (Example4_7), digits = 2)
install.packages ("psych")
library (psych)
install.packages ("GPArotation")
library (GPArotation)
corr.test (Example4_7)
```

Example 4.8

```
round (cov (Example4_8), digits = 3)
round (cor (Example4_8), digits = 3)
install.packages ("psych")
library (psych)
install.packages ("GPArotation")
library (GPArotation)
corr.test (Example4_8)
```

Example 4.9

```
jpeg ("FIGURE 4.1.jpeg", res = 400, height = 8, width = 8, units = 'cm')
par (mar = c(5, 2, 3, 1))
plot (Example4_9 $ PV, Example4_9 $ AV, xlab = "", ylab = "",
cex = 0.5, pch = 10, col = "blue", cex.axis = 0.5, xaxt = "n", yaxt ="n")
axis (1, c(1, 5, 10, 15, 20), cex.axis = 0.5, tck = -0.03, mgp = c(3, .3, 0))
axis (2, c(2.5, 3, 3.5, 4), cex.axis = 0.5, tck = -0.03, mgp = c(3, .3, 0))
mtext (c("PV"), side = 1, line = 1, cex = 0.5)
mtext (c("AV"), side = 2, line = 1, cex = 0.5)
dev.off()
```

Figure 4.2

```
#Figure4_2a
jpeg ("Figure4_2a.jpeg", res = 400,height = 8, width = 8, units = 'cm')
par (mar = c(2.5, 2, 1, 1))
plot (Figure4_2a$x, Figure4_2a$y, xlab = "", ylab = "", cex = 0.5, pch = 19,
col = "blue", cex.axis = 0.5, tck = -0.03, mgp = c (3, 0.3, 0), xaxt = "n", yaxt = "n")
axis (1, seq(1, 10, 2), cex.axis = 0.5, tck = -0.03, mgp = c(3, .3, 0))
axis (2, seq(0, 100, 20), cex.axis = 0.5, tck = -0.03, mgp = c(3, .3, 0))
mtext (c("x"), side = 1, line = 1, cex = 0.5)
mtext (c("Y"), side = 2, line = 1, cex = 0.5)
dev.off()
#Figure4_2b
jpeg ("Figure4_2b.jpeg", res = 400, height = 8, width = 8, units = 'cm')
Par (mar = c(2.5, 2, 1, 1))
Plot (Figure4_2b$x, Figure4_2b$y, xlab = "", ylab = "", cex = 0.5, pch = 19, col = "blue", cex.axis = 0.5,
tck = -0.03, mgp = c(3, .3, 0), xaxt = "n", yaxt = "n")
Axis (1, seq(0, 12, 2), cex.axis = 0.5, tck = -0.03, mgp = c(3, .3, 0))
Axis (2, seq(40, 100, 10), cex.axis = 0.5, tck = -0.03, mgp = c(3, .3, 0))
mtext (c("x"), side = 1, line = 1, cex = 0.5)
mtext (c ("Y"), side = 2, line = 1, cex = 0.5)
dev.off()
```

Example 4.10

```
jpeg ("FIGURE 4.3.jpeg", res = 400, height = 8, width = 10, units = 'cm')
par(mar = c(1, 1, 1, 1))
pairs (~ PV + PAV + AV + TBA, data = Example4_10,col = "blue", cex = 0.5, pch = 10, cex.axis = 0.7, tck =
-0.04, mgp = c(3, .3, 0))
dev.off()
```

Example 4.11

```
round (dist (Example4_11), digits = 2)
```

Example 4.12

```
round (dist (Example4_12), digits = 2)
```

Chapter 5

Example 5.2

```
#Example 5.2 (a)
alpha = 0.05
z1 = round (qnorm (1 - alpha/2) , digits = 2) # critical value
c.vs = c(-z1, z1)
```

(Continued)

(Continued)

```
c.vs          #Pring critical values
x = seq (-3, 3, 0.1)          # graph the function
y = dnorm (x)
jpeg ("FIGURE 5.1a.jpeg", res = 400, height = 8, width = 8, units = 'cm')
par (mar = c (5, 5, 3, 1))
plot (x, y, type = "l", col = "black", xaxt = "n", xlab = "", cex = 0.5, ylab = "Probability")
x1 = seq (-3, -z1, length = 200)
y1 = dnorm (x1)
polygon (c(-3, x1, -z1), c(0, y1, 0), col = "gray")
axis (1, at = -z1, col.axis = "black", cex.axis = 0.5, tck = -0.03, mgp = c (3, .3, 0))
arrows (-2.5, 0.1, -2.5, 0.001, xpd = FALSE, length = 0.05)
text (-2.5, 0.12, alpha/2, col = "blue", cex = 0.5)
text (0, 0.07,"non-rejection region", cex = 0.5)
x2 = seq (2,3,length = 200)
y2 = dnorm(x2)
polygon (c(z1, x2, 3), c(0, y2, 0), col = "gray")
axis (1, at = z1, col.axis = "black", cex.axis = 0.5, tck = -0.03, mgp = c(3, .3, 0))
mtext (c("Z"), side = 1, line = 1, cex = 0.7)
axis (1, at = 0, col.axis = "black", cex.axis = 0.5, tck = -0.03, mgp = c(3, .3, 0))
arrows (2.5, 0.1, 2.5, 0.001, xpd = FALSE, length = 0.05)
text (2.5, 0.12, alpha/2, col = "blue", cex = 0.5)
dev.off()


#Example 5.2 (b)
alpha = 0.05
z1 = round (qnorm (1 - alpha) , digits = 3)
z1
c.vs = c(-z1, z1)
c.vs     #Pring critical values
x = seq (-3, 3, 0.1)
y = dnorm (x)
jpeg ("FIGURE 5.1b.jpeg", res = 400, height = 8, width = 8, units = 'cm')
par (mar = c (5, 5, 3, 1))
plot (x, y, type = "l", col = "black", xaxt = "n", xlab = "", cex = 0.5, ylab = "Probability")
text (-0.1, 0.05,"non-rejection region", cex = 0.5)
x2 = seq (z1,3,length = 200)
y2 = dnorm (x2)
polygon (c (z1, x2, 3), c (0, y2, 0), col = "gray")
axis (1, at = z1, col.axis = "black", cex.axis = 0.5, tck = -0.03, mgp = c(3, .3, 0))
mtext (c("Z"), side = 1, line = 1, cex = 0.7)
axis (1, at = 0, col.axis = "black", cex.axis = 0.5, tck = -0.03, mgp = c(3, .3, 0))
arrows (2, 0.1, 2, 0.001, xpd = FALSE, length = 0.05)
text (2, 0.12, alpha, col = "blue", cex = 0.5)
dev.off()


#Example 5.2 (c)
alpha = 0.05
z1 = round (qnorm (1 - alpha) , digits = 3)
z1
c.vs = c(-z1, z1)
```

(Continued)

```
c.vs
x = seq (-3, 3, 0.1)                # graph the function
y = dnorm (x)
jpeg ("FIGURE 5.1c.jpeg", res = 400, height = 8, width = 8, units = 'cm')
par (mar = c(5, 5, 3, 1))
plot (x, y, type = "l", col = "black", xaxt = "n", xlab = "", ylab = "Probability")
text (0.2, 0.03, "non-rejection region", cex = 0.5)
x2 = seq(-z1,-3,length = 200)
y2 = dnorm (x2)
polygon (c(-z1, x2, -3), c(0, y2, 0), col = "gray")
axis (1, at = -z1, col.axis = "black", cex.axis = 0.5, tck = -0.03, mgp = c(3, .3, 0))
mtext (c("Z"), side = 1, line = 1, cex = 0.7)
axis (1, at = 0, col.axis = "black", cex.axis = 0.5, tck = -0.03, mgp = c(3, .3, 0))
arrows (-2, 0.1, -2, 0.001, xpd = FALSE, length = 0.05)
text (-2, 0.12, alpha, col = "blue", cex = 0.5)
dev.off()
```

Example 5.3

```
mean = 5.11                    # sample mean
mu = 6                         # hypothesized value
s = 0.44                       # population standard deviation
n = 36                         # sample size
z = (mean - mu)/( s / sqrt(n))
z                              # print z-test statistic
alpha = 0.05
z1 = round (qnorm (1 - alpha/2) , digits = 2) # critical value
c.vs = c(-z1, z1)                   # critical value for two sided
c.vs                          #Pring critical values
x = seq (-3, 3, 0.1)                # graph the function
y = dnorm (x)
jpeg("FIGURE 5.2.jpeg", res = 400, height = 8, width = 8, units = 'cm')
par (mar = c(4.5, 2, 3, 1))
plot (x, y, type = "l", col = "black", xaxt = "n", xlab = "", yaxt = "n")
x1 = seq (-3,-z1,length = 200)
y1 = dnorm(x1)
polygon (c(-3, x1, -z1), c(0, y1, 0), col = "gray")
axis (1, at = -z1, col.axis = "black", cex.axis = 0.5, tck = -0.03, mgp = c(3, .3, 0))
arrows (-2.3, 0.1, -2.3, 0.001, xpd = FALSE, length = 0.05)
text (-2.2, 0.12, alpha/2, col = "blue", cex = 0.5)
text (0, 0.07, "non-rejection region", cex = 0.5)
x2 = seq (2, 3, length = 200)
y2 = dnorm (x2)
polygon (c(z1, x2, 3), c(0, y2, 0), col = "gray")
axis (1, at = z1, col.axis = "black", cex.axis = 0.5, tck = -0.03, mgp = c(3, .3, 0))
axis(2, seq(0,1,0.1), cex.axis = 0.5, tck = -0.03, mgp = c(3, .3, 0))
arrows (2.4, 0.1, 2.4, 0.001, xpd = FALSE, length = 0.05)
text(2.5, 0.12, alpha/2, col = "blue", cex =0.5)
text (-2.6, 0.17, round (z , digits = 2), col = "blue", cex = 0.5)
arrows (-2.8, 0.15, -2.8, 0.001, xpd = FALSE, length = 0.05)
```

(Continued)

(Continued)

```
mtext (side = 1, "Z", line = 0.5, cex = 0.5 )
mtext (c ("Probability"), side = 2, line = 1, cex = 0.5)
dev.off()
```

Example 5.4

```
mean = 27                       # sample mean
mu = 19                         # hypothesized value
s = 5                           # standard deviation
n = 42                          # sample size
z = (mean - mu)/(s / sqrt(n))
z                               # print z-test statistic
alpha = 0.01
z1 = round (qnorm (1 - alpha/2) , digits = 2) # critical value
c.vs = c(-z1, z1)          # critical value for two sided
c.vs                            #Pring critical values
x = seq (-3, 3, 0.1)          # graph the function
y = dnorm (x)
jpeg ("FIGURE 5.3.jpeg", res = 400, height = 8, width = 8, units = 'cm')
par (mar = c(4, 2, 3, 1))
plot (x, y, type = "l", col = "black", xaxt = "n", yaxt = "n")
x1 = seq(-3,-z1, length = 200)
y1 = dnorm (x1)
polygon (c (-3, x1, -z1), c(0, y1, 0),col = "gray")
axis (1, at = -z1, col.axis = "black", cex.axis = 0.5, tck = -0.03, mgp=c(3, .3, 0))
arrows (-2.7, 0.1, -2.7, 0.001, xpd = FALSE, length = 0.05)
text (-2.6, 0.12, alpha/2, col = "blue", cex = 0.5)
text (0, 0.07, "non-rejection region", cex = 0.5)
x2 = seq (z1, 3, length = 200)
y2 = dnorm (x2)
polygon (c(z1, x2, 3), c(0, y2, 0), col = "gray")
axis (1, at = z1, col.axis = "black", cex.axis = 0.5, tck = -0.03, mgp = c (3, .3, 0))
arrows (2.7, 0.1, 2.7, 0.001, xpd = FALSE, length = 0.05)
text (2.4, 0.12, alpha/2, col = "blue", cex = 0.5)
text (2.7, 0.17, round (z, digits = 2), col = "blue", cex = 0.5)
arrows (3, 0.15, 3, 0.001, xpd = FALSE, length = 0.05)
mtext (side = 1, "Z", line = 0.2, cex = 0.5 )
axis (2, seq(0,1,0.1), cex.axis = 0.5, tck = -0.03, mgp = c(3, .3, 0))
mtext (c("Probability"), side = 2, line = 1, cex = 0.5)
dev.off()
```

Figure 5.4 t distribution curves

```
x <- seq(-5, 5, length = 200)
y <- dnorm(x)
d.f <- c(1, 5, 50, 100)
colors <- c ("red", "blue", "green", "black", "gray")
```

(Continued)

(Continued)

```
labels <- c ("d.f = 1", "d.f = 5", "d.f = 50","d.f = 100", "normal")
jpeg ("FIGURE 5.4.jpeg", res = 400, height = 8, width = 8, units = 'cm')
par(mar = c(5, 3, 3, 1))
plot(x, y, type = "l", lty = 2,ylab = "", xlab = "", pch = 19, cex = 0.5, xaxt = "n", yaxt = "n")
for (i in 1:6){
  lines(x, dt(x,d.f[i]), lwd = 1, col = colors[i])
}
Legend ("topright", inset = .01, labels, lwd = 2, lty = c(2, 2, 2, 2), cex = 0.5, col = colors)
Axis (1, seq (-5,5, 2), cex.axis = 0.5, tck = -0.03, mgp = c(3, .3, 0))
Axis (2, seq (0, 1, 0.1), cex.axis = 0.5, tck = -0.03, mgp = c(3, .3, 0))
mtext (side = 2, "Probability", line = 1, cex = 0.5 )
mtext (side = 1, "X value", line = 1, cex = 0.5 )
dev.off()
```

Example 5.5

```
#Two-sided t-test
mean = 0.23                      # sample mean
mu = 0.15                        # hypothesized value
s = 0.02                         # standard deviation
n = 10                           # sample size
t = (mean - mu) / (s / sqrt (n))
t                                # print t-test statistic
alpha = 0.05
c.v = qt (1 - alpha/2, df = n-1)      # critical value
t1 = round (c.v, digits = 2)
c.vs = c(-t1, t1)                # critical value for two sided
c.vs                             #Pring critical values
x = seq (-3, 3, 0.1)             # graph the function
y = dnorm (x)
jpeg ("FIGURE 5.5.jpeg", res = 400, height = 8, width = 8, units = 'cm')
par (mar = c(2.5, 2, 1, 1))
plot (x, y, type = "l", col = "black", xaxt = "n", xlab = "", yaxt = "n")
x1 = seq (-3, -t1,length = 200)
y1 = dnorm (x1)
polygon (c(-3, x1, -t1), c(0, y1, 0),col = "gray")
axis (1, at = -t1, col.axis = "black", cex.axis = 0.5, tck = -0.03, mgp = c(3, .3, 0))
arrows (-2.7, 0.1, -2.7, 0.001, xpd = FALSE, length = 0.05)
text (-2.7, 0.12, alpha/2, col = "blue", cex = 0.5)
text (0, 0.07, "non-rejection region", cex = 0.5)
x2 = seq (t1, 3,length = 200)
y2 = dnorm (x2)
polygon (c(t1, x2, 3), c(0, y2, 0), col = "gray")
axis (1, at = t1, col.axis = "black", cex.axis = 0.5, tck = -0.03, mgp = c(3, .3, 0))
arrows (2.7, 0.1, 2.7, 0.001, xpd = FALSE, length = 0.05)
text (2.7, 0.12, alpha/2, col = "blue", cex = 0.5)
text (2.9, 0.17, round (t, digits = 2), col = "blue", cex = 0.5)
arrows (3, 0.15, 3, 0.001, xpd = FALSE, length = 0.05)
mtext (side = 1, "Z", line = 0.2, cex = 0.5 )
```

(Continued)

(Continued)

```
axis (2, seq(0, 1, 0.1), cex.axis = 0.5, tck = -0.03, mgp = c(3, .3, 0))
mtext (c("Probability"), side = 2, line = 1, cex = 0.5)
dev.off()
```

Example 5.6

```
mean = 3.2                      # sample mean
mu = 5                          # hypothesized value
s = 1.13                        # standard deviation
n = 21                          # sample size
t = (mean - mu)/( s/ sqrt(n))
t                   # print t-test statistic
alpha = 0.10
c.v = qt (1 - alpha/2, df = n-1)      # critical value
t1 = round (c.v, digits = 2)
c.vs = c(-t1, t1)                     # critical value for two sided
c.vs                                  #Pring critical values
x = seq (-3, 3, 0.1)                  # graph the function
y = dnorm (x)
jpeg ("FIGURE 5.6.jpeg", res = 400, height = 8, width = 8, units = 'cm')
par (mar = c(4.5, 2, 3, 1))
plot (x, y, type = "l", col = "black", xaxt = "n", yaxt = "n", xlab = "", ylab = "")
x1 = seq (-3, -t1, length = 200)
y1 = dnorm (x1)
polygon (c(-3, x1, -t1), c (0, y1, 0), col = "gray")
axis (1, at = -t1, col.axis = "black", tck = -0.03, mgp = c(3, .3, 0), cex.axis = 0.5)
arrows (-2.7, 0.1, -2.7, 0.001, xpd = FALSE, length = 0.05)
text (-2.7, 0.12, alpha/2, col = "blue", cex = 0.5)
text (0, 0.07, "non-rejection region", cex = 0.5)
x2 = seq (t1, 3, length = 200)
y2 = dnorm(x2)
polygon (c (t1, x2, 3), c (0, y2, 0), col = "gray")
axis (1, at = t1, col.axis = "black", tck = -0.03, mgp = c(3, .3, 0), cex.axis = 0.5)
arrows (2.7, 0.1, 2.7, 0.001, xpd = FALSE, length = 0.05)
text (2.7, 0.12, alpha/2, col = "blue", cex = 0.5)
text (-2.9, 0.17, round (t, digits = 2), col = "blue", cex = 0.5)
arrows (-3, 0.15, -3, 0.001, xpd = FALSE, length = 0.05)
mtext(side = 1, "Z", line = 0.2, cex = 0.5 )
axis(2, seq(0, 1, 0.1), cex.axis = 0.5, tck = -0.03, mgp = c(3, .3, 0), cex =0.5)
mtext (c("Probability"), side = 2, line = 1, cex = 0.5)
dev.off()
```

Example 5.7

```
t.test (Example5_7)
t.test (Example5_7, mu = 0.10)
t.test (Example5_7, mu = 0.10, alternative = "greater")
t.test (Example5_7, mu = 0.10, alternative = "less")
```

Example 5.8

```
M <- round (sapply (Example5_8, mean), digits = 3)
M                           # print the mean vector
Mn <- c (7, 49, 73)         # Hypothesized mean vector
d = M - Mn                  # the difference between the mean and claim
d
dt = t (d)                  #transpose for the difference
dt
n <- length (Example5_8 $ pH)    #number of observations
n
co = round (cov (Example5_8), digits = 3)    # covariance matrix
co
IN = solve (co)             # calculate the inverse
IN
HOT = dt % * % IN % * % d * n
HOT                         #Hotelling's value
```

Example 5.9

```
M <- round (sapply (Example5_9, mean),digits = 3)
M                           # print the mean vector
Mn <- c (19, 4)             # Hypothesized mean vector
d = M - Mn                  # the difference between the mean and claim
d
dt = t(d)                   #transpose for the difference
dt
n <- length (Example5_9 $ Fat.content)    #number of observations
n
co = round (cov (Example5_9), digits = 3)    # covariance matrix
co
IN = solve (co)             # calculate the inverse
IN
HOT = dt % * % IN % * % d * n
HOT                         # Hotelling's value
```

Example 5.10

```
T <- t.test (Example5_10 $ Honey, Example5_10 $ Zehdi)
T
T$statistic
#Calculate the critical values for the graph
alpha = 0.05
n1 = length (Example5_10 $ Honey)
n2 = length (Example5_10 $ Zehdi)
c.v = qt (1-alpha/2, df = n1 + n2 - 2)    #critical value
t1 = round (c.v, digits = 3)
t1
c.vs = c(-t1, t1)
```

(Continued)

(Continued)

```
c.vs
x = seq (-3, 3, 0.1)            # graph the function
y = dnorm (x)
jpeg ("Figure 5.7.jpeg", res = 400, height = 8, width = 8, units = 'cm')
par( mar = c(4.5, 2.5, 3, 1))
plot (x, y, type = "l", col = "black", xaxt = "n", yaxt = "n", ylab = "", xlab = "")
x1 = seq (-3, -t1, length = 200)
y1 = dnorm (x1)
polygon (c (-3, x1, -t1), c (0, y1, 0), col = "gray")
axis (1, at = -t1, col.axis = "black", cex.axis = 0.5, tck = -0.03, mgp = c(3, .3, 0))
arrows (-2.7, 0.1, -2.7, 0.001, xpd = FALSE, length = 0.05)
text (-2.7, 0.12, alpha/2, col = "blue", cex = 0.5)
text (0, 0.07,"non-rejection region", cex = 0.5)
x2 = seq (t1, 3,length = 200)
y2 = dnorm (x2)
polygon (c (t1, x2, 3), c (0, y2, 0), col = "gray")
axis (1, at = t1, col.axis = "black", cex.axis = 0.5, tck = -0.03, mgp = c(3, .3, 0))
arrows (2.7, 0.1, 2.7, 0.001, xpd = FALSE, length = 0.05)
text (2.7, 0.12, alpha/2, col = "blue", cex = 0.5)
text (-2.9, 0.17, round (T $ statistic, digits = 2), col = "blue", cex = 0.5)
arrows (-3, 0.15, -3, 0.001, xpd = FALSE, length = 0.05)
mtext (side = 1, "Z", line = 0.2, cex = 0.5 )
axis (2, seq(0, 1, 0.1), cex.axis = 0.5, tck = -0.03, mgp = c(3, .3, 0))
mtext (c("Probability"), side = 2, line = 1, cex = 0.5)
dev.off()
```

Example 5.11

```
T <- t.test (Example5_11 $ Cavendish, Example5_11 $ Dream)
T
T $ statistic
# calculate the critical value
n1 = length (Example5_11 $ Cavendish)
n2 = length (Example5_11 $ Dream)
df = n1 + n2 - 2
alpha = 0.05
t1 = round (qt(1 - alpha/2, df = n1 + n2 - 2), digits = 3)
t1
c.vs = c(-t1, t1)
x = seq (-3, 3, 0.1)              # graph the function
y = dnorm(x)
jpeg ("Figure 5.8.jpeg", res = 400, height = 8, width = 8, units = 'cm')
par ( mar = c(4.5, 2.5, 3, 1))
plot (x, y, type = "l", col = "black", xaxt = "n", yaxt = "n", ylab = "", xlab = "")
x1 = seq (-3, -t1, length = 200)
y1 = dnorm(x1)
polygon (c (-3, x1, -t1), c (0, y1, 0), col = "gray")
axis (1, at = -t1, col.axis = "black", cex.axis = 0.5, tck = -0.03, mgp = c (3, .3, 0))
arrows (-2.7, 0.1, -2.7, 0.001, xpd = FALSE, length = 0.05)
```

(Continued)

(Continued)

```
text (-2.7, 0.12, alpha/2, col = "blue", cex = 0.5)
text (0, 0.07, "non-rejection region", cex = 0.5)
x2 = seq (t1, 3, length = 200)
y2 = dnorm (x2)
polygon (c(t1, x2, 3), c(0, y2, 0), col = "gray")
axis (1, at = t1, col.axis = "black", cex.axis = 0.5, tck = -0.03, mgp = c (3, .3, 0))
arrows (2.7, 0.1, 2.7, 0.001, xpd = FALSE, length = 0.05)
text (2.7, 0.12, alpha/2, col = "blue", cex = 0.5)
text (-0.53, 0.17, round (T $ statistic, digits = 2), col = "blue", cex = 0.5)
arrows (-0.52, 0.15, -0.52, 0.01, xpd = FALSE, length = 0.05)
mtext (side = 1, "Z", line = 0.2, cex = 0.5 )
axis (2, seq(0, 1, 0.1), cex.axis = 0.5, tck = -0.03, mgp = c(3, .3, 0))
mtext (c ("Probability"), side = 2, line = 1, cex = 0.5)
dev.off()
```

Example 5.12

```
m1 <- sapply (Example5_12a, mean)
m1
m2 <- sapply (Example5_12b, mean)
m2
round (cov (Example5_12a), digits = 3)
round (cov (Example5_12b), digits = 3)
# calculate Sp
n1 = length (Example5_12a $ Cu)
n2 = length (Example5_12b $ Cu)
sp = (((n1 - 1) * cov (Example5_12a)) + ((n2 - 1) * cov (Example5_12b))) / (n1 + n2 - 2)
round (sp, digits = 3)
dif = m1 - m2          # calculate the difference
dif1 = t (dif)        # transpose
INV = solve (sp)      # calculate the inverse
HOT = ((n1 * n2) / (n1 + n2)) * (dif1 % * % INV % * % dif)
HOT
```

Example 5.13

```
#calculate the mean vector
m1 <- sapply (Example5_13a, mean)
m1
m2 <- sapply (Example5_13b, mean)
m2
round (cov (Example5_13a), digits = 3)
round (cov (Example5_13b), digits = 3)
# calculate Sp
n1 = length (Example5_13a $ pH)
n2 = length (Example5_13b $ pH)
sp = (((n1 - 1) * cov(Example5_13a)) + ((n2 - 1) * cov (Example5_13b))) / (n1 + n2 - 2)
```

(Continued)

(Continued)

```
round (sp, digits = 3)
dif = m1 - m2        # calculate the difference
dif1 = t(dif)    # transpose
INV = solve (sp)  # calculate the inverse
TSQ = ((n1 * n2) / (n1 + n2)) * (dif1 % * % INV % * % dif)
TSQ
```

Chapter 6

Example 6.1

```
av < - aov(GI ~ Type.of.Noodles, data = Example6_1)
av
summary (av)
aov.out = aov (GI ~ Type.of.Noodles, data = Example6_1)
aov.out
summary (aov.out)
```

Example 6.2

```
av < - aov (TFC ~ Type.of.date, data = Example6_2)
av
summary (av)
aov.out = aov (TFC ~ Type.of.date, data = Example6_2)
aov.out
summary (aov.out)
```

Example 6.3

```
Snack = as.factor (Example6_3 $ Snack)
Snack
Packaging = as.factor (Example6_3 $ packaging)
Peroxide = Example6_3 $ Peroxide
aov (Peroxide ~ Snack*Packaging)      # ANOVA with interaction
summary (aov(Peroxide ~ Snack*Packaging))
jpeg("Figure 6.1.jpeg", res = 400, height = 8, width = 8, units = 'cm')
par( mar = c(5, 3, 3, 1))
interaction.plot (Snack, Packaging, Peroxide, col = c("yellow", "blue", "black"),legend = F, ylab = "" ,
xlab = "", cex.axis = .5, tck = -0.03, mgp = c(3, .3, 0))
legend ("topleft", c("p","L", "t"),lwd = 2, pch = c(18, 18, 18 ),
     col = c("yellow","blue", "black"),inset = .01, lty = c(2, 2, 2),cex = 0.5)
mtext (side = 1, "Snack", line = 0.2, cex = 0.5 )
mtext (side = 2, "Mean of peroxide", line = 1.5, cex = 0.5 )
dev.off()
```

Example 6.4

```
Temperature = factor (Example6_4 $ Temperature, levels = c(50, 60, 70))
Acid = factor (Example6_4 $ Acid, levels = c("Oxalic", "Citric", "Combine"))
Weight = Example6_4 $ Weight
aov (Weight ~ Temperature*Acid)        # ANOVA with interaction
summary (aov(Weight ~ Temperature*Acid))
jpeg("Figure 6.2.jpeg", res = 400, height = 8, width = 8, units = 'cm')
par( mar = c(5, 3.5, 3, 2))
interaction.plot (Temperature, Acid, Weight, col = c("blue", "yellow", "black"), legend = F, ylab = "" ,
xlab = "", cex.axis = .5, tck = -0.03, mgp = c(3, .3, 0))
legend("topright", c("Oxalic", "Citric", "Combine"), lty = c(1,3,4), cex = .3, col = c("blue", "yellow",
"black"))
axis (1, at = c(50, 60, 70), line = 0.2, col.axis = "black", tck = -0.03, mgp = c(3, .3, 0), cex.axis = 0.5, )
mtext(side = 1, "Temperature", line = 0.2, cex = 0.5 )
mtext(side = 2, "Mean of weight", line = 1.5, cex = 0.5 )
dev.off()
```

Example 6.5

```
TS = Example6_5 $ TS
Elasticity = Example6_5 $ Elasticity
Time = as.factor (Example6_5 $ Time)
Y <- cbind (TS, Elasticity)
Ma <- manova (Y ~ Time)
summary (Ma, test = "Wilks")
summary (Ma, test = "Hotelling-Lawley")
summary (Ma, test = "Roy")
summary (Ma, test = "Pillai")
```

Example 6.6

```
Brand = as.factor (Example6_6 $ Brand)
Y <- cbind (Example6_6 $ Fat.content, Example6_6 $ moisture, Example6_6 $ L, Example6_6 $ a, Example6_6 $ b)
Ma <- manova (Y ~ Brand)
summary (Ma, "Wilks")
summary (Ma, "Hotelling-Lawley")
summary (Ma, "Pillai")
summary (Ma, "Roy")
```

Example 6.7

```
Cd = Example6_7 $ Cd
Pb = Example6_7 $ Pb
Zn = Example6_7 $ Zn
Cu = Example6_7 $ Cu
Ni = Example6_7 $ Ni
Y <- cbind (Cd, Pb, Zn, Cu, Ni)
```

(Continued)

(Continued)

```
Type = as.factor (Example6_7 $ Type)
Con. = as.factor (Example6_7 $ Con.)
Ma <- manova (Y ~ Type*Con.)
summary (Ma, test = "Wilks")
summary (Ma, test = "Hotelling-Lawley")
summary (Ma, test = "Pillai")
summary (Ma, test = "Roy")
```

Example 6.8

```
pH = Example6_8 $ pH
PV = Example6_8 $ PV
TBA = Example6_8 $ TBA
Y <- cbind (pH, PV, TBA)
Types_dates = as.factor (Example6_8 $ Typesofdates)
concentration = as.factor (Example6_8 $ concentration)
Ma <- manova (Y ~ Types_dates*concentration)
summary (Ma, test = "Pillai")
summary (Ma, test = "Hotelling-Lawley")
summary (Ma, test = "Roy")
summary (Ma, test = "Wilks")
```

Chapter 7

Example 7.1

```
model = lm (y ~ x, data = Example7_1)
model
anova (lm (y ~ x, data = Example7_1))
summary (lm (y ~ x, data = Example7_1))
```

Example 7.2

```
model = lm (Cd ~ Cu, data = Example7_2)
model
anova (lm (Cd ~ Cu, data = Example7_2))
summary (lm (Cd ~ Cu, data = Example7_2))
```

Example 7.3

```
model = (lm (y ~ x, data = Example7_1))
newdata = data.frame (x = 27)
predict (model, newdata, interval = "predict")
```

Example 7.4

```
model1 = lm (Cd1 ~ Cu1, data = Example7_2a) #Before changing
model1
Cu1 = Example7_2a$Cu1
Cd1 = Example7_2a$Cd1
jpeg ("FIGURE 7.1a.jpeg", res = 400, height = 8, width = 8, units = 'cm')
Par (mar = c(7, 4, 3, 1.5))
plot (Cu1 ,Cd1 , pch =19, col = "blue", cex = .5)  ##before
abline (model1, col = "black")  #draw regression line for before
dev.off()
model = lm (Cd ~ Cu, data = Example7_4)     # After changing
model
Cu = Example7_4$Cu
Cd = Example7_4$Cd
jpeg ("FIGURE 7.1b.jpeg", res = 400, height = 8, width = 8, units = 'cm')
par(mar = c(7, 4, 3, 1.5))
plot (Cu, Cd, pch =19, col = "blue", cex = .5)  ##after
abline (model, col = "black")  #draw regression line for after
dev.off()
```

Example 7.5

```
model = lm (y ~ x, data = Example7_5)
model
anova (lm (y ~ x, data = Example7_5))
summary (lm(y ~ x, data = Example7_5))
predicted = round (predict(model), digits = 3)
predicted
round (resid(model), digits = 3)
M = mean (Example7_5 $ y)
round ((predicted - M)^2, digits = 3)
sum ((predicted - M)^2)          #Explained
round ((Example7_5 $ y - predicted)^2, digits = 3)
sum ((Example7_5 $ y - predicted)^2)      #Unexplained
```

Example 7.6

```
Coefficient of determination can be obtained from ANOVA.
```

Example 7.7

```
model < - lm (y ~ x1 + x2, data = Example7_7)
model
anova (lm (y ~ x1 + x2, data = Example7_7)) # anova table
summary (lm (y ~ x1 + x2, data = Example7_7)) # show results
```

Chapter 8

Example 8.1

```
install.packages ("psych")
library (psych)
install.packages ("GPArotation")
library (GPArotation)
# calculate the variance
round (sapply (Example8_1[2:14], var), digits = 2)
corr.test (Example8_1[,2:14])
#principal component analysis
pca < - princomp (Example8_1[2:14], cor = TRUE)
summary (pca) # print variance accounted for
#Figure 8.1
jpeg ("Figure 8.1.jpeg", res = 400, height = 8, width = 8, units = 'cm')
Par (mar = c (4.5,2,6,1.5))
screeplot(pca, npcs = 13,type = "barplot", xaxt = "n", yaxt = "n")
mtext (c("Variances"), side = 2, line = 1, cex = 0.7)
mtext (c("Components"), side = 1, line = 0.3, cex = 0.7)
axis (2, seq(0, 6, 1), cex.axis = 0.5, tck = -0.03, mgp = c(3, .3, 0))
dev.off()
round (unclass (pca $ loadings), digits = 3) #To get the loadings
round (pca $ scores, digits = 3) # the principal components
##Bar plot for the scores
A = as.data.frame (pca $ scores)
#Figure 8.2
#A[,1] #represent the scores for the first principal component
jpeg ("Figure 8.2.jpeg", res = 400, height = 8, width = 8, units = 'cm')
par(mar = c(4.5, 3.5, 3, 1.5))
barplot (A[,1], xaxt = "n", yaxt = "n", ylim = c(-5, 5), col = c (rep ('gray', 12), rep ('darkkhaki', 12),
rep ('orange', 12), rep ('blue', 12)))
legend (45, 5, horiz = FALSE, c ("Gpe", "Rpe", "Gpu", "Rpu"),inset = c(0,.2), col = c ("gray", "darkkhaki",
"orange", "blue"), pch = c(19, 19, 19, 19), cex = 0.7)
mtext (c ("Scores of PC1"), side = 2, line = 2, cex = 0.7)
mtext (c ("Variety of bananas"), side = 1, line = 0.2, cex = 0.7)
axis (2, seq(-5, 5, 1), cex.axis = 0.5, tck = -0.03, mgp = c(3, .3, 0))
dev.off()
#Figure 8.3
jpeg ("Figure 8.3.jpeg", res = 400, height = 8, width = 8, units = 'cm')
par(mar = c (4.5, 3.5, 3, 1.5))
barplot (A[,2], ylim = c(-6, 6), xaxt = "n", yaxt = "n", col = c (rep ('gray', 12), rep ('darkkhaki', 12),
rep ('orange', 12), rep ('blue', 12)))
legend (45, 6, horiz = FALSE, c("Gpe", "Rpe", "Gpu", "Rpu"),
     col = c("gray", "darkkhaki", "orange", "blue"), pch = c(19, 19, 19, 19), cex = 0.7)
axis (2, seq(-5, 5, 1), cex.axis = 0.5, tck = -0.03, mgp = c(3, .3, 0))
mtext (c("Scores of PC2"), side = 2, line = 2, cex = 0.75)
mtext (c("Variety of bananas"), side = 1, line = 0.2, cex = 0.75)
dev.off()
#Figure 8.4
jpeg ("Figure 8.4.jpeg", res = 400, height = 8, width = 8, units = 'cm')
Par (mar = c(4.5, 3.5, 3, 1.5))
```

(Continued)

(Continued)

```r
barplot (A[,3], ylim = c(-3, 3), xaxt ="n", yaxt ="n", col = c (rep ('gray', 12), rep ('darkkhaki', 12),
rep ('orange', 12), rep ('blue', 12)))
legend (45, 3, horiz = FALSE, c ("Gpe", "Rpe", "Gpu", "Rpu"),
    col = c ("gray", "darkkhaki", "orange", "blue"), pch = c(19, 19, 19, 19), cex = 0.5)
axis(2, seq(-5, 5, 1), cex.axis = 0.5, tck = -0.03, mgp = c (3, .3, 0))
mtext (c ("Scores for PC3"), side = 2, line = 2, cex = 0.75)
mtext (c ("Variety of bananas"), side = 1, line = 0.2, cex = 0.75)
dev.off()
#Figure 8.5
jpeg ("Figure 8.5.jpeg", res = 400, height = 8, width = 8, units = 'cm')
par(mar = c (4.5, 2.5, 3, 1.5))
plot (A[,1], A[,2], xlab = "", ylab ='', pch = 19, ylim = c(-4, 4), xaxt ="n", yaxt ="n", col = c (rep
('gray', 12), rep ('darkkhaki', 12), rep ('orange', 12), rep ('blue', 12)))
abline (h = 0, v = 0)
legend ("topright", horiz = FALSE, c("Gpe", "Rpe", "Gpu", "Rpu"), col = c("gray", "darkkhaki", "orange",
"blue"), pch = c(19, 19, 19, 19), cex = 0.5)
axis (2, seq(-4, 4, 1), cex.axis = 0.5, tck = -0.03, mgp = c(3, .3, 0))
axis (1, seq(-4, 4, 1), cex.axis = 0.5, tck = -0.03, mgp = c(3, .3, 0))
mtext (c("Scores for the second component"), side = 2, line = 1, cex = 0.5)
mtext (c("Scores for the first component"), side = 1, line = 1, cex = 0.5)
dev.off()
```

Example 8.2

```r
install.packages ("psych")
library (psych)
install.packages ("GPArotation")
library (GPArotation)
## Calculate the variance
round (sapply (Example8_2[2:5], var), digits = 2)
corr.test (Example8_2[2:5])
## calculate PCA
pca <- princomp (Example8_2[2:5], cor = TRUE)
summary(pca) # print variance accounted for
jpeg ("Figure 8.6.jpeg", res = 400, height = 8, width = 8, units = 'cm')
par(mar = c (5, 2.2, 6, 1))
screeplot (pca, xaxt ="n", yaxt ="n") # scree plot  xlim = c (1,15)
mtext (c ("Variances"), side = 2, line = 1, cex = 0.5)
mtext (c ("Components"), side = 1, line = 0.75, cex = 0.5)
axis (2, seq(0, 5, 1), cex.axis = 0.5, tck = -0.03, mgp = c(3, .3, 0))
axis (1, seq(1, 5, 1), cex.axis = 0.5, tck = -0.03, mgp = c(3, .3, 0))
dev.off()
round (unclass (pca $ loadings), digits = 2)
round (pca $ scores, digits = 3)
#Bar plot for the scores
A = as.data.frame (pca $ scores)
A[,1]
#Type = as.factor (Example8_2$Type)
#Group
```

(Continued)

```
jpeg ("Figure 8.7.jpeg", res = 400, height = 8, width = 8, units = 'cm')
par( mar = c(4.5, 2.5, 3.5, 1.5))
barplot (A[,1], xlab = "", ylim = c(-6, 6),ylab ='', pch = 19, xaxt ="n", yaxt ="n", col = c (rep
('gray', 3), rep ('darkgreen', 3), rep ('orange', 3),rep ('red', 3), rep ('blue', 3), rep ('yellow', 3),
rep ('darkkhaki', 3), rep ('black',3)))
legend ("topright", horiz = FALSE, c("Honey", "Shahroon", "Bam", "Jiroft", 'Piarom', 'Kabkab', 'Zahedi',
'Kharak'), col = c ("gray", "darkgreen", "orange", "red", 'blue',"yellow", 'darkkhaki', 'black'), pch = c
(19, 19, 19, 19, 19, 19, 19, 19), cex = .40)
axis (2, seq(-4, 5, 1), cex.axis = 0.5, tck = -0.03, mgp = c(3, .3, 0))
mtext (c("Scores for the PC1"), side = 2, line = 1, cex = 0.5)
mtext (c("Type of dates"), side = 1, line = 0.75, cex = 0.5)
dev.off()
```

Chapter 9

```
Example 9.1
```

```
install.packages ("psych")
library (psych)
install.packages ("GPArotation")
library (GPArotation)
corr.test (Example9_1[,2:10])
PCA <- princomp (Example9_1[,2:10],cor = TRUE)
summary (PCA)
#Figure 9.1
jpeg ("Figure 9.1.jpeg", res = 400, height = 8, width = 8, units = 'cm')
par(mar = c(4, 2, 6, 0.05))
screeplot (PCA , npcs = 10, type = "barplot", xaxt ="n", yaxt ="n", ylim = c(0,4))
mtext (c("Variances"), side = 2, line = 1, cex = 0.70)
mtext (c("Components"), side = 1, line = 0.3, cex = 0.70)
axis(2, seq(0, 4, .5), cex.axis = 0.5, tck = -0.03, mgp = c(3, .3, 0))
dev.off()
round (unclass (PCA $ loadings), digits = 2)
Factor <- principal (Example9_1[,2:10], nfactors = 4, rotate = "varimax", covar = FALSE)
Print (Factor, digits = 2)
Round (Factor $ scores, digits = 2)
#Figure 9.2
jpeg ("Figure 9.2.jpeg", res = 400, height = 8, width = 8, units = 'cm')
par (mar = c(4, 2.2, 3, 1))
barplot (Factor $ scores[,1], ylim = c(-2, 2) , xaxt ="n", yaxt ="n",pch = 19, col = c(rep ('gray', 11),
rep ('orange', 11)))
legend (18, 2, horiz = FALSE, c("Cavendish", "Dream"),inset = c(0,.2), col = c('gray', 'orange'), pch = c
(19, 19), cex = 0.50)
mtext (c("Scores for the first factor"), side = 2, line = 1, cex = 0.70)
mtext (c("Variety of bananas"), side = 1, line = 0.2, cex = 0.70)
axis (2, seq(-2, 2, 1), cex.axis = 0.5, tck = -0.03, mgp = c(3, .3, 0))
dev.off()
#Figure 9.3
jpeg ("Figure 9.3.jpeg", res = 400, height = 8, width = 8, units = 'cm')
par (mar = c(4, 2.2, 3, 1))
```

(Continued)

(Continued)

```
barplot (Factor $ scores[,2], ylim = c(-3,3), xaxt ="n", yaxt ="n", pch = 19, col = c (rep ('gray', 11),
rep ('orange', 11)))
legend (18, 3, horiz = FALSE, c("Cavendish", "Dream"),inset = c(0,.2), col = c('gray', 'orange'), pch = c
(19, 19), cex = 0.50)
mtext (c("Scores for the second factor"), side = 2, line = 1, cex = 0.7)
mtext (c("Variety of bananas"), side = 1, line = 0.2, cex = 0.7)
axis(2, seq(-3, 3, 1), cex.axis = 0.5, tck = -0.03, mgp = c(3, .3, 0))
dev.off()
#Figure 9.4
jpeg ("Figure 9.4.jpeg", res = 400, height = 8, width = 8, units = 'cm')
par(mar = c (4, 2, 3, 1))
barplot (Factor $ scores[,3],ylim = c(-2,2), xaxt ="n", yaxt ="n", pch = 19, col = c(rep ('gray', 11),rep
('orange', 11)))
legend (18, 2, horiz = FALSE, c("Cavendish", "Dream"),inset = c(0,.2), col = c('gray', 'orange'), pch = c
(19, 19), cex = 0.50)
mtext (c("Scores for the third factor"), side = 2, line = 1, cex = 0.7)
mtext (c("Variety of bananas"), side = 1, line = 0.2, cex = 0.7)
axis (2, seq(-2, 3, 1), cex.axis = 0.5, tck = -0.03, mgp = c(3, .3, 0))
dev.off()
#Figure 9.5
jpeg ("Figure 9.5.jpeg", res = 400, height = 8, width = 8, units = 'cm')
par (mar = c (4, 2.5, 3, 1))
barplot (Factor $ scores[,4],ylim = c(-3,4), xaxt ="n", yaxt ="n", pch = 19, col = c(rep('gray',11),rep
('orange',11)))
legend (18, 3, horiz = FALSE, c("Cavendish", "Dream"),inset = c(0,.2), col = c('gray', 'orange'), pch = c
(19, 19), cex = 0.50)
mtext (c("Scores for the fourth factor"), side = 2, line = 1, cex = 0.7)
mtext (c("Variety of bananas"), side = 1, line = 0.2, cex = 0.7)
axis (2, seq(-3, 3, 1), cex.axis = 0.5, tck = -0.03, mgp = c(3, .3, 0))
dev.off()
#Figure 9.6
# Draw the first and second principal component
jpeg ("Figure 9.6.jpeg", res = 400, height = 8, width = 8, units = 'cm')
par(mar = c(5, 2, 3, 1))
plot (Factor $ scores[, 1], Factor $ scores[, 2] , ylab = '', xlab = '', xaxt ="n", yaxt ="n", xlim=c
(-3,3) , ylim=c(-3,3), pch = 19, cex = 0.50, col = c(rep("gray",11), rep("orange",11)))
abline (h = 0,v = 0)
legend (1,3, horiz = FALSE, c("Cavendish", "Dream"),inset= c(0,.2), col = c('gray', 'orange'), pch = c
(19, 19), cex = 0.50)
mtext (c("Scores for the second factor"), side = 2, line = 1, cex = 0.7)
mtext (c("Scores for the first factor"), side = 1, line = 1, cex = 0.7)
axis (2, seq(-3, 4, 1), cex.axis = 0.5, tck = -0.03, mgp = c(3, .3, 0))
axis(1, seq(-3, 4, 1), cex.axis = 0.5, tck = -0.03, mgp = c(3, .3, 0))
dev.off()
```

Example 9.2

```
install.packages ('psych')
library (psych)
```

(Continued)

(Continued)

```
install.packages ('GPArotation')
library (GPArotation)
corr.test (Example9_2[,2:13])
PCA <- princomp (Example9_2[,2:13],cor = TRUE)
summary (PCA)
#Figure 9.7
jpeg ("Figure 9.7.jpeg", res = 400, height = 8, width = 8, units = 'cm')
par (mar = c(4, 2, 6, 1.5))
screeplot (PCA , npcs = 10, type = "barplot", xaxt = "n", yaxt = "n", ylim = c(0, 6))
mtext (c("Variances"), side = 2, line = 1, cex = 0.7)
mtext (c("Components"), side = 1, line = 0.3, cex = 0.7)
axis(2, seq(0, 6, 1), cex.axis = 0.5, tck = -0.03, mgp = c(3, .3, 0))
dev.off()
round (unclass (PCA $ loadings),digits = 2)
Factor <- principal (Example9_2[,2:13], nfactors = 3, rotate = "varimax")
Factor
round (Factor $ scores, digits = 2)
#Figure 9.8
jpeg ("Figure 9.8.jpeg", res = 400, height = 8, width = 8, units = 'cm')
par(mar = c(4, 2.2, 3, 1))
barplot (Factor $ scores[,1], ylim = c(-2,2) , xaxt = "n", yaxt = "n",pch = 19, col = c(rep ('gray', 3), rep
('darkgreen', 3), rep ('orange', 3), rep ('red', 3), rep ('blue', 3), rep ('yellow', 3), rep ('darkkhaki',
3)))
legend (22, 2, horiz = FALSE, c('T1', 'T2', 'T3', 'T4', 'T5', 'T6', 'T7'),inset = c(0,.2), col = c('gray',
'darkgreen', 'orange', 'red', 'blue', 'yellow', 'darkkhaki'), pch = c(19, 19), cex = 0.50)
mtext (c ("Scores for the first factor"), side = 2, line = 1, cex = 0.7)
mtext (c("Type of tapioca"), side = 1, line = 0.2, cex = 0.7)
axis (2, seq(-2, 2, 1), cex.axis = 0.5, tck = -0.03, mgp = c(3, .3, 0))
dev.off()
#Figure 9.9
jpeg ("Figure 9.9.jpeg", res = 400, height = 8, width = 8, units = 'cm')
par ( mar = c(4, 2.2, 3, 1))
barplot (Factor $ scores[,2], ylim = c(-2,2) , xaxt = "n", yaxt = "n",pch = 19, col = c(rep ('gray', 3),rep
('darkgreen', 3),rep ('orange', 3), rep ('red', 3), rep ('blue', 3), rep ('yellow', 3), rep ('darkkhaki', 3)))
legend (22, 2, horiz = FALSE, c('T1', 'T2', 'T3', 'T4', 'T5', 'T6', 'T7'),inset = c(0,.2), col = c('gray',
'darkgreen', 'orange', 'red', 'blue', 'yellow', 'darkkhaki'), pch = c(19, 19), cex = 0.60)
mtext (c("Scores for the second factor"), side = 2, line = 1, cex = 0.7)
mtext (c("Type of tapioca"), side = 1, line = 0.2, cex = 0.7)
axis (2, seq(-2, 2, 1), cex.axis = 0.5, tck = -0.03, mgp = c(3, .3, 0))
dev.off()
#Figure9.10
jpeg ("Figure 9.10.jpeg", res = 400, height = 8, width = 8, units = 'cm')
par (mar = c(4, 2.2, 3, 1))
barplot (Factor $ scores[,3], ylim = c(-2,2) , xaxt = "n", yaxt = "n", pch = 19, col = c(rep ('gray', 3),
rep ('darkgreen', 3),rep ('orange', 3), rep ('red', 3), rep ('blue', 3), rep ('yellow', 3), rep ('darkkha-
ki', 3)))
legend (.5, 2, horiz = TRUE, c('T1', 'T2', 'T3', 'T4', 'T5', 'T6', 'T7'),inset = c(0,.2), col = c('gray',
'darkgreen', 'orange', 'red', 'blue', 'yellow', 'darkkhaki'), pch = c(19, 19), cex = 0.45)
mtext (c("Scores for the second factor"), side = 2, line = 1, cex = 0.7)
mtext (c("Type of tapioca"), side = 1, line = 0.2, cex = 0.7)
axis (2, seq(-2, 2, 1), cex.axis = 0.5, tck = -0.03, mgp = c(3, .3, 0))
```

(Continued)

(Continued)

```
dev.off()
# Draw the first and second factors
jpeg ("Figure 9.11.jpeg", res = 400, height = 8, width = 8, units = 'cm')
par(mar = c(5, 2.2, 3, 1))
plot(Factor$scores[,1],Factor$scores[,2], xaxt = "n", yaxt = "n", pch = 19, ylab = '', xlab = '', ylim = c
(-2.5, 2.5), xlim = c(-2, 2), col = c(rep ('gray', 3),rep ('burlywood', 3),rep ('orange', 3),rep ('red',
3), rep('blue', 3),rep('yellow', 3),rep('black', 3)))
legend (1.5, 2.5, horiz = FALSE, c('T1', 'T2', 'T3', 'T4', 'T5', 'T6', 'T7'),inset = c(0,.2), col = c
('gray', 'burlywood', 'orange', 'red', 'blue', 'yellow', 'black'), pch = c(19, 19), cex = 0.40)
abline (h = 0, v = 0)
mtext (c("Scores for the second factor"), side = 2, line = 1, cex = 0.5)
mtext (c("Scores for the first factor"), side = 1, line = 1, cex = 0.5)
axis (2, seq(-2, 2, 0.5), cex.axis = 0.5, tck = -0.03, mgp = c(3, .3, 0))
axis (1, seq(-2, 2, 0.5), cex.axis = 0.5, tck = -0.03, mgp = c(3, .3, 0))
dev.off()
```

Chapter 10

Example 10.1

```
install.packages ("psych")
library (psych)
install.packages ("GPArotation")
library (GPArotation)
install.packages ("MASS")
library (MASS)
DA <- lda (Location ~ Cr + As + Cd + Zn + Cu + Pb + Hg, data = Example10_1)
DA
Scores <- predict (DA)
Scores      # scores for each sample
B = as.data.frame (Scores)
#Figure 10.1
jpeg ("Figure 10.1.jpeg", res = 600, height = 8, width = 8, units = 'cm')
par(mar = c (4, 2.2, 3, 1))
barplot (B $ LD1, ylim = c(-4,4), xaxt = "n", yaxt = "n", pch = 19, col = c (rep ('gray', 20), rep ('blue', 20)))
legend (36, 4, horiz = FALSE, c('Juru River', 'Jejawi River'),inset = c(0,.2), col = c('gray', 'blue'),
pch = c(19, 19), cex = 0.50)
mtext (c("Scores for the first discriminant function"), side = 2, line = 1, cex = 0.6)
mtext (c ("Sites"), side = 1, line = 0.2, cex = 0.7)
axis (2, seq(-4, 4, 1), cex.axis = 0.5, tck = -0.03, mgp = c (3, .3, 0))
dev.off()
```

Example 10.2

```
cT <- table (Example10_1 $ Location, Scores $ class)       #classification Table
cT
prop.table (cT, 1)
diag (prop.table (cT, 1))    #correct Classification percentage
sum (diag (prop.table (cT)))  # total percent correct
```

Example 10.3

```
Library (psych)
Library (GPArotation)
Library (MASS)
DA <- lda (Brand ~ FC + Mo. + L + a + b + PV + PAV + AV + TBA + SAF + MUFA + PUFA, data = Example10_3)
DA
Scores <- predict (DA)
Scores      # scores for each sample
B = as.data.frame (Scores)
#B
#B$x.LD1
jpeg ("Figure 10.2.jpeg", res = 400, height = 8, width = 8, units = 'cm')
par(mar = c (4, 2.2, 3, 1))
barplot (B$x.LD1, xaxt = "n", yaxt = "n",pch = 19, col = c (rep ('gray', 3), rep ('darkgreen', 3), rep
('orange', 3), rep ('red', 3), rep ('blue', 3), rep ('yellow', 3), rep ('darkkhaki', 3)))
legend (22, 10500, horiz = FALSE, c('T1', 'T2', 'T3', 'T4', 'T5', 'T6', 'T7'),inset = c(0,.2), col = c
('gray', 'darkgreen', 'orange', 'red', 'blue', 'yellow', 'darkkhaki'), pch = c(19, 19), cex = 0.50)
mtext (c("Scores for the first discriminant function"), side = 2, line = 1, cex = 0.6)
mtext (c("Brand of Tapioca chips"), side = 1, line = 0.2, cex = 0.7)
axis (2, seq(-6000, 12000, 1000), cex.axis = 0.5, tck = -0.03, mgp = c(3, .3, 0))
dev.off()
#Figure 10.3
jpeg ("Figure 10.3.jpeg", res = 400, height = 8, width = 8, units = 'cm')
par(mar = c (4, 2.2, 3, 1))
barplot (B$x.LD2, xaxt = "n", yaxt = "n",pch = 19, ylim = c(-3000, 4000), col = c (rep ('gray', 3), rep
('darkgreen', 3), rep ('orange', 3), rep ('red', 3), rep ('blue', 3), rep ('yellow', 3), rep ('darkkhaki', 3)))
legend (1, 4000, horiz = TRUE, c('T1', 'T2', 'T3', 'T4', 'T5', 'T6', 'T7'),inset = c(0,.2), col = c
('gray', 'darkgreen', 'orange', 'red', 'blue', 'yellow', 'darkkhaki'), pch = c(19, 19), cex = 0.43)
mtext (c("Scores for the first discriminant function"), side = 2, line = 1, cex = 0.5)
mtext (c("Brand of Tapioca chips"), side = 1, line = 0.2, cex = 0.7)
axis (2, seq(-3000, 4000, 1000), cex.axis = 0.5, tck = -0.03, mgp = c(3, .3, 0))
dev.off()
#first and second
jpeg ("Figure 10.4.jpeg", res = 400, height = 8, width = 8, units = 'cm')
par(mar = c (5, 2.2, 3, 1))
plot (B$x.LD1, B$x.LD2, xaxt = "n", yaxt = "n", pch = 19,
    ylim = c(-4000, 4000), xlim = c(-11000, 12000), cex = .6, col = c (rep ('gray', 3), rep ('burlywood',
3), rep ('orange', 3), rep ('red', 3), rep ('blue', 3), rep ('yellow', 3), rep ('black', 3)))
legend (10000, 4000, horiz = FALSE, c('T1', 'T2', 'T3', 'T4', 'T5', 'T6', 'T7'),inset = c(0,.2), col = c
('gray', 'burlywood', 'orange', 'red', 'blue', 'yellow', 'black'), pch = c(19, 19,19, 19,19,19, 19), cex
= 0.50)
abline (h = 0, v = 0)
mtext (c("Scores for the second discriminant function"), side = 2, line = 1, cex = 0.5)
mtext (c("Scores for the first discriminant function"), side = 1, line = 1, cex = 0.5)
axis (2, seq(-4000, 4000, 1000), cex.axis = 0.5, tck = -0.03, mgp = c(3, .3, 0))
axis (1, seq(-11000, 12000, 1000), cex.axis = 0.5, tck = -0.03, mgp = c(3, .3, 0))
dev.off()
```

Example 10.4

```
#Classification
cT <- table (Example10_3 $ Brand, Scores $ class)        #classification Table
cT
prop.table (cT, 1)      # correct and misclassification percentage
diag (prop.table (cT, 1))      #correct Classification percentage
sum (diag (prop.table (cT))) # total percent correct
```

Chapter 11

Figure 11.1

```
#a. single
Distance <- round (dist(Figure11_1[2:8]),digits = 2)
Distance
hc <- hclust (Distance, method = "single")       # apply hierarchical clustering
hc
jpeg("Figure 11.1a.jpeg", res = 400,height = 12, width = 10, units = 'cm')
par(mar = c(8, 6, 6, 3))
plot(hc, labels = as.character(Figure11_1$Site),font.main = 1, cex.lab = 0.5, cex.main = .5,font.sub =
1, cex.sub = .5, cex.axis = 0.5, cex = 0.5) # plot the dendrogram
#plot(hc, labels = as.character(Figure11_1$Site), main = NULL, sub = "", xlab = "")  # plot the dendrogram
without main and sub titles
dev.off()
#b. Average
Distance <- round(dist(Figure11_1[2:8]),digits = 2)
Distance
hc <- hclust(Distance, method = "average")         # apply hierarchical clustering
hc
jpeg("Figure 11.1b.jpeg", res = 400,height = 12, width = 10, units = 'cm')
par(mar = c(8, 6, 6, 3))
plot(hc, labels = as.character(Figure11_1$Site),font.main = 1, cex.lab = 0.5, cex.main = .5,font.sub =
1, cex.sub = .5, cex.axis = 0.5, cex = 0.5)  # plot the dendrogram
dev.off()
#c. complete
Distance <- round(dist(Figure11_1[2:8]),digits = 2)
Distance
hc <- hclust(Distance,method = "complete")            # apply hierarchical clustering
hc
jpeg("Figure 11.1c.jpeg", res = 400,height = 12, width = 10, units = 'cm')
par(mar = c(8, 6, 6, 3))
plot(hc, labels = as.character(Figure11_1$Site),font.main = 1, cex.lab = 0.5, cex.main = .5,font.sub =
1, cex.sub = .5, cex.axis = 0.5, cex = 0.5)  # plot the dendrogram
dev.off()
```

Example 11.1

```
Distance <- round (dist (Example11_1[2:7]), digits = 2)  # find distance matrix
Distance
```

(*Continued*)

(Continued)

```
CA <- hclust (Distance, method = "single")        # apply hierarchical clustering
CA
jpeg ("Figure 11.2.jpeg", res = 400,height = 12, width = 10, units = 'cm')
par ( mar = c (8, 6, 6, 3))
plot (CA,labels = Example11_1 $ Variety, font.main = 1, cex.lab = 0.5, cex.main = .5, font.sub = 1, cex.sub
= .5, cex.axis = 0.5, cex = 0.5)   # plot the dendrogram
dev.off()
```

Example 11.2

```
Distance <- round (dist (Example11_2[2:13]),digits = 2)
Distance
CA <- hclust (Distance,"single")
CA
jpeg ("Figure 11.3.jpeg", res = 1000,height = 12, width = 10, units = 'cm')
Par ( mar = c (8, 6, 6, 3))
Plot (CA, labels = Example11_2 $ Type,font.main = 1, cex.lab = 0.5, cex.main = .5,font.sub = 1, cex.sub =
.5, cex.axis = 0.5, cex = 0.5)
dev.off()
```

Index

Note: Page numbers followed by "*f*" and "*t*" refer to figures and tables, respectively.

Printed in the United States
By Bookmasters